Highway and Rail Transit Tunnel Inspection Manual

2005 Edition

U.S. Department of Transportation

Federal Highway Administration
Federal Transit Administration

Published by Books Express Publishing, 2012
ISBN 978-1-78266-164-1

Books Express publications are available from all good retail and online booksellers. For publishing proposals and direct ordering please contact us at: info@books-express.com

TABLE OF CONTENTS

Table of Contents
List of Tables
List of Figures
Executive Summary

CHAPTER 1: INTRODUCTION 1-1

CHAPTER 2: TUNNEL CONSTRUCTION AND SYSTEMS 2-1

    A. Tunnel Types 2-1
        1. Shapes 2-1
        2. Liner Types 2-6
        3. Invert Types 2-7
        4. Construction Methods 2-9
        5. Tunnel Finishes 2-11
    B. Ventilation Systems 2-13
        1. Types 2-13
        2. Equipment 2-17
    C. Lighting Systems 2-18
        1. Types 2-18
    D. Other Systems/Appurtenances 2-19
        1. Track 2-19
        2. Power (Third Rail/Catenary) 2-20
        3. Signal/Communication Systems 2-22

CHAPTER 3: FUNDAMENTALS OF TUNNEL INSPECTION 3-1

    A. Inspector Qualifications 3-1
        1. Civil/Structural 3-1
        2. Mechanical 3-2
        3. Electrical 3-2
        4. Track, Third Rail, Catenary, Signals and Communications 3-4
    B. Responsibilities 3-5
    C. Equipment/Tools 3-5
    D. Preparation 3-6
        1. Mobilization 3-6
        2. Survey Control 3-7
        3. Inspection Forms 3-11
    E. Methods of Access 3-23

    F. Safety Practices 3-23
        1. Highway 3-23
        2. Rail Transit 3-23

CHAPTER 4: INSPECTION PROCEDURES – GENERAL DISCUSSION 4-1

    A. Inspection of Civil/Structural Elements 4-1
        1. Frequency 4-1
        2. What to Look For 4-1
        3. Safety – Critical Repairs 4-11
        4. Condition Codes 4-11
        5. Tunnel Segments 4-12
    B. Inspection of Mechanical Systems 4-18
        1. Frequency 4-18
        2. What to Look For 4-19
    C. Inspection of Electrical Systems 4-21
        1. Frequency 4-21
        2. What to Look For 4-22
    D. Inspection of Other Systems/Appurtenances 4-25
        1. Inspection of Track Elements 4-25
        2. Inspection of Power Systems (Third Rail/Catenary) 4-29
        3. Inspection of Signal/Communication Systems 4-34

CHAPTER 5: INSPECTION DOCUMENTATION 5-1

    A. Field Data 5-1
        1. Tunnel Structure 5-1
        2. Track Structure 5-2
        3. Specialized Testing Reports 5-2
    B. Repair Priority Definitions 5-6
        1. Critical 5-6
        2. Priority 5-6
        3. Routine 5-6
    C. Reports 5-6

Glossary G-1
References R-1

LIST OF TABLES

Table 2.1 – Construction Methods 2-9

Table 3.1 – Liner Type Acronyms 3-12

Table 4.1 – General Condition Codes 4-12

Table 4.2 – Condition Code Summary 4-15

Table 4.3 – Mechanical Inspection Frequency 4-18

Table 4.4 – Electrical Inspection Frequency 4-22

Table 4.5 – Passenger Train Operating Speeds 4-25

Table 4.6 – Track Gage Distances 4-27

Table 4.7 – Track Alignment 4-27

LIST OF FIGURES

Figure 2.1 – Circular Highway Tunnel Shape 2-1

Figure 2.2 – Double Box Highway Tunnel Shape 2-2

Figure 2.3 – Horseshoe Highway Tunnel Shape 2-2

Figure 2.4 – Oval/Egg Highway Tunnel Shape 2-3

Figure 2.5 – Circular Rail Transit Tunnel Shape 2-3

Figure 2.6 – Double Box Rail Transit Tunnel Shape 2-4

Figure 2.7 – Single Box Rail Transit Tunnel Shape 2-4

Figure 2.8 – Horseshoe Rail Transit Tunnel Shape 2-5

Figure 2.9 – Oval Rail Transit Tunnel Shape 2-5

Figure 2.10 – Circular Tunnel Invert Type 2-8

Figure 2.11 – Single Box Tunnel Invert Type 2-8

Figure 2.12 – Horseshoe Tunnel Invert Type 2-9

Figure 2.13 – Natural Ventilation 2-14

Figure 2.14 – Longitudinal Ventilation 2-14

Figure 2.15 – Semi-Transverse Ventilation 2-15

Figure 2.16 – Full-Transverse Ventilation 2-16

Figure 2.17 – Axial Fans 2-17

Figure 2.18 – Centrifugal Fan 2-17

Figure 2.19 – Typical Third Rail Power System 2-21

Figure 2.20 – Typical Third Rail Insulated Anchor Arm 2-21

Figure 3.1 – Tunnel Inspection Layout Plan 3-8

Figure 3.2 – Circular Tunnel Clock System Designations 3-9

Figure 3.3 – Circular Tunnel Label System Designations 3-9

Figure 3.4 – Rectangular Tunnel Label System Designations 3-10

Figure 3.5 – Horseshoe Tunnel Label System Designations 3-10

Figure 5.1 – Tunnel Inspection Form (Tablet PC Data Collector) 5-3

Figure 5.2 – Tunnel Inspection Form (Pre-Printed Form) 5-4

Figure 5.3 – Portal Inspection Form (Pre-Printed Form) 5-5

EXECUTIVE SUMMARY

In March of 2001, the Federal Highway Administration (FHWA), in conjunction with the Federal Transit Administration (FTA), engaged Gannett Fleming, Inc., to develop the first ever Tunnel Management System to benefit both highway and rail transit tunnel owners throughout the United States and Puerto Rico. Specifically, these federal agencies, acting as ONE DOT, set a common goal to provide uniformity and consistency in assessing the physical condition of the various tunnel components. It is commonly understood that numerous tunnels in the United States are more than 50 years old and are beginning to show signs of considerable deterioration, especially due to water infiltration. In addition, it is desired that good maintenance and rehabilitation practices be presented that would aid tunnel owners in the repair of identified deficiencies. To accomplish these ONE DOT goals, Gannett Fleming, Inc., was tasked to produce an Inspection Manual, a Maintenance and Rehabilitation Manual, and a computerized database wherein all inventory, inspection, and repair data could be collected and stored for historical purposes.

This manual provides specific information for the inspection of both highway and rail transit tunnels. Although several components are similar in both types of tunnels, a few elements are specific to either highway or rail transit tunnels and are defined accordingly. The following paragraphs explain the specific subjects covered along with procedural recommendations that are contained in this manual.

Introduction

This chapter presents a brief history of the project development and outlines the scope and contents of the Inspection Manual.

Tunnel Construction and Systems

To develop uniformity concerning certain tunnel components and systems, this chapter was developed to define those major systems and describe how they relate to both highway and rail transit tunnels. This chapter is broken down into four sub-chapters, which include: tunnel types, ventilation systems, lighting systems, and other systems/appurtenances.

The tunnel types section covers the different tunnel shapes in existence, liner types that have been used, the two main invert types, the various construction methods used to construct a tunnel, and the multiple different finishes that typically exist in highway tunnels. The ventilation and lighting system sections are self explanatory in that they cover the basic system types and configurations. The other systems/appurtenances section is used to explain tunnel systems that are present in rail transit tunnels, such as: track systems, power systems (third rail/ catenary), and signal/ communications systems.

Fundamentals of Tunnel Inspection

As can be expected, there are basic steps that must be properly accomplished for the end product of the inspection to be useful to the tunnel owner for planning purposes. These steps include making sure that the inspectors are qualified to properly identify defects and make recommendations about their respective systems within the tunnel. Also, the responsibilities of the individual inspection team members and the tunnel owner are discussed. The next section lists the equipment/tools that may be required to perform the inspections.

A section on preparation for the inspection consists of describing the tasks that should be completed during the mobilization phase of the inspection. Also, a survey control section is given that describes how to record the inspection results with respect to their location within the tunnel. Following that, suggested standard forms are presented that can be used to record the actual structural condition codes assessed during the inspection.

After the preparation section, brief sections on methods of access, which describes equipment that might be necessary to reach the areas that need to be inspected, and safety practices for both highway and rail transit tunnels are included.

Inspection Procedures – General Discussion

This chapter presents recommended frequencies and specific defects to look for in each of the following categories: structural elements, mechanical systems, electrical systems, and other systems/appurtenances.

The structural elements section includes descriptions of defects in concrete, steel, masonry, and timber. Also included in this section is a segment describing the procedures that should be followed in the event that the inspection reveals defects that require immediate repair. Structural conditions codes are detailed on a 0 to 9 scale for the general condition and subsequently for specific tunnel segments for cut-and-cover box tunnels, soft ground tunnel liners, rock tunnel liners, and timber liners. The individual tunnel segment ratings are summarized in a table.

The systems/appurtenances section includes general discussions on track elements, power systems (third rail/catenary), and signal/communication systems. Given the complexity of these systems, only general inspection recommendations are given for the major components.

Inspection Documentation

The final chapter of this manual offers suggestions on how to properly record the results of an in-depth inspection. The field data section describes how to visually record the defects that are found, either on pre-printed forms or through the use of tablet PC's (pen based computers) in to a database. Abbreviations are given for the most common defects that are found on the tunnel structure. Also included are recommendations that are specific to the track structure and any specialized testing reports that were generated during the inspection.

Repair priority definitions are presented so that the individuals writing the inspection report can classify the defects based on definitions for critical, priority, and routine classifications. Finally, a recommended outline for the inspection report is given for guidance as to what information should be included for the tunnel owners' use in determining how to address the items identified during the inspection.

CHAPTER 1:
INTRODUCTION

Background

The National Bridge Inspection Standards (NBIS) were established in the early 1970s to ensure highway structures received proper inspection using uniform procedures and techniques. The NBIS address a number of issues including personnel qualification, inspection frequency, and reporting of inspection findings. Following the issuance of the NBIS, the Federal Highway Administration (FHWA) developed a comprehensive training course, including an inspector's manual, designed for those individuals in the highway community responsible for bridge inspection. The training course and manual covered the typical types of highway structures in the nation, providing information on inspection procedures for the various components of those structures. Missing from the material was appropriate procedures to employ for preparing and conducting inspections on the various features of highway tunnels. Tunnels were considered unique structures and special applications would be needed for them.

Recently, the FHWA created an office specifically to focus on management of highway assets. This office has a major function—to work with the highway community to design, develop, and implement state-of-the-art systems for managing highway assets, including bridges and pavements. One area of the highway needing emphasis was a management system for tunnels. Similarly, the Federal Transit Administration (FTA) is responsible for providing transit tunnel owners with a wide range of assistance, including guidance on appropriate management techniques. Because of the common interest on tunnel management procedures from both agencies, the FHWA and FTA have joined to sponsor the development of a system for highway and rail transit tunnels. A project to develop the system was initiated in March of 2001 to include preparing an inventory of highway and rail transit tunnels in the U.S., an inspection manual, a manual for maintenance and repair, and a computer software program for data management. All of these products will be furnished to each highway and transit tunnel owner across the nation, and will be available as public domain.

Scope

The purpose of this manual is to provide highway and rail transit tunnel owners guidance for establishing procedures and practices for the inspection, documentation, and priority classification of deficiencies for various elements that comprise an existing tunnel. It is also the intent that this manual be used as part of a comprehensive inspection and maintenance program. The preliminary research performed indicates that a majority of tunnel owners believe there is a need to develop guidance for procedures for managing tunnel activities that could be readily implemented.

This manual addresses inspection procedures for the functional aspect of the tunnel, focusing on the civil/structural, mechanical, and electrical components. The manual does, however, provide brief guidance on other systems/appurtenances, such as track, traction power, signals, and communications, which comprise the operational aspects of a rail transit tunnel. This brief guidance is only meant to provide general knowledge and not in-depth inspection criteria for such systems/appurtenances.

Contents

To ensure consistency of definition of particular elements, this manual contains several chapters that explain the various types of elements that exist within the tunnel. For example, the description of tunnel components such as tunnel configuration, liner types, invert types, ventilation systems, lighting systems, tunnel finishes and other systems/appurtenances (track, traction power, signals, and communications) are each provided in separate sections to assist tunnel owners in educating their inspectors as to the particular system used for the tunnel to be inspected. Furthermore, the manual provides suggested guidelines for inspection personnel qualifications and equipment to be used for performing the inspection. Since most tunnels are constructed of concrete, steel, masonry, and timber (to a very limited degree), this manual provides extensive definitions of the types of common defects that occur within these major structural elements so that the inspection documentation is consistent according to the guidelines provided.

The manual contains procedures for documenting the inspection findings. These range from identifying a particular defect (cracking, scaling, spalling, corrosion, etc.) and its severity (minor, moderate, or severe) to assessing the overall condition of an element within a particular region of the tunnel. The manual is based upon a condition assessment scale that varies from "0" to "9," with 0 being the worst condition and 9 being the best condition. This is similar to the scale used for the National Bridge Inventory that is familiar to most highway/transit tunnel owners. The length of a tunnel segment for which these ratings will be applied will vary with each tunnel and tunnel owner. Some tunnels have panels that are numbered between particular joints, which make it easy for determining the segment length over which condition assessments are to be evaluated. Other tunnel owners may choose to have the evaluation performed for a segment of a tunnel, say 30 m (100 ft) or 60 m (200 ft). Regardless, the entire tunnel is to be inspected and condition assessments applied for all tunnel segments.

The manual will also provide guidance for the inspector to prioritize defects for repair and rehabilitation. Although this manual proposes the use of three prioritizations for conducting repairs, namely critical, priority, and routine, tunnel owners can adopt other prioritizations as appropriate.

This manual is developed for a hands-on, up-close inspection of the tunnel structure. The procedures developed herein are for visual and non-destructive methods of evaluating the tunnel elements. This does not preclude the lead inspector from requesting that certain destructive means (e.g., extracting cores for determination of freeze/thaw resistance or concrete strength) be requested to aid in determining soundness/adequacy of the tunnel elements.

Although this manual is produced for a hands-on, non-destructive evaluation of the inside face of the tunnel structure, other state-of-the-art, non-destructive testing methods may be used in areas that require a more in-depth structural evaluation. These methods may include mechanical oscillation techniques such as sonic or ultrasonic measurements (more commonly referred to as Impact-Echo), electronic techniques such as georadar, and optical techniques such as infrared thermography and multispectral analysis. Each of the above methods has been used successfully in tunnels; however, a full understanding of the applications and limitations of each method is necessary to maximize their benefits.

It is felt, however, that these state-of-the-art methods are probably only cost effective in long, rail transit tunnels in metropolitan areas. It is assumed that these methods will mostly supplement and not replace the hands-on, non-destructive testing methods described in this manual for many tunnel owners in the United States.

CHAPTER 2:
TUNNEL CONSTRUCTION AND SYSTEMS

A. TUNNEL TYPES

This section describes the various types of highway and rail transit tunnels. These tunnel types are described by their shape, liner type, invert type, construction method, and tunnel finishes. It should be noted that other types may exist currently or be constructed in the future as new technologies become available. The purpose of this section is to look at the types that are most commonly used in tunnel construction to help the inspector properly classify any given tunnel. As a general guideline a minimum length of 100 meters (~300 feet) was used in defining a tunnel for inventory purposes. This length is primarily to exclude long underpasses, however other reasons for using the tunnel classification may exist such as the presence of lighting or a ventilation system, which could override the length limitation.

1. Shapes

a) Highway Tunnels

As shown in Figures 2.1 to 2.4, there are four main shapes of highway tunnels – circular, rectangular, horseshoe, and oval/egg. The different shapes typically relate to the method of construction and the ground conditions in which they were constructed. Although many tunnels will appear rectangular from inside, due to horizontal roadways and ceiling slabs, the outside shape of the tunnel defines its type. Some tunnels may be constructed using combinations of these types due to different soil conditions along the length of the tunnel. Another possible highway tunnel shape that is not shown is a single box with bi-directional traffic.

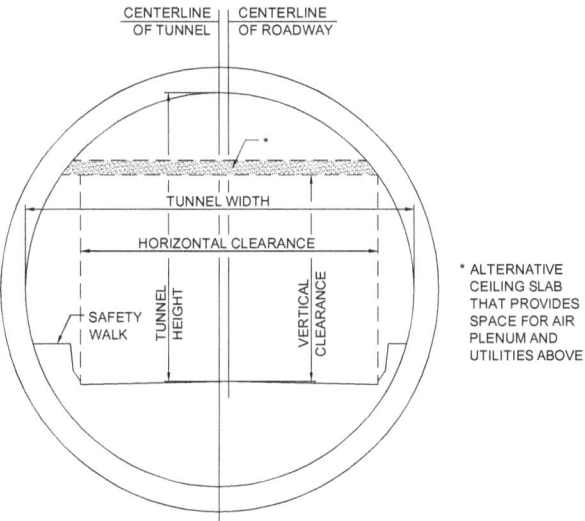

Figure 2.1 – Circular tunnel with two traffic lanes and one safety walk. Also shown is an alternative ceiling slab. Invert may be solid concrete over liner or a structural slab.

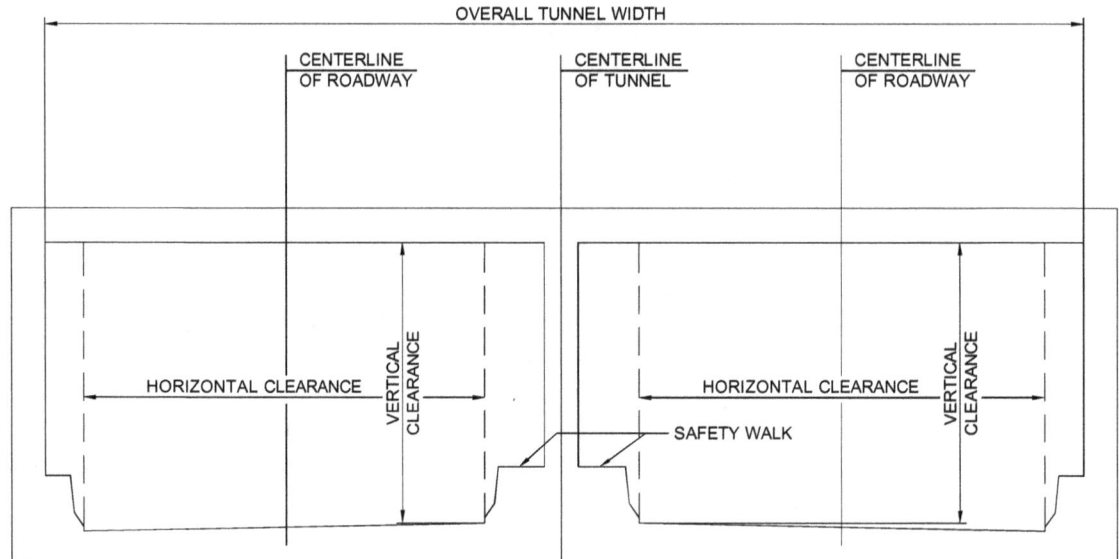

Figure 2.2 – Double box tunnel with two traffic lanes and one safety walk in each box. Depending on location and loading conditions, center wall may be solid or composed of consecutive columns.

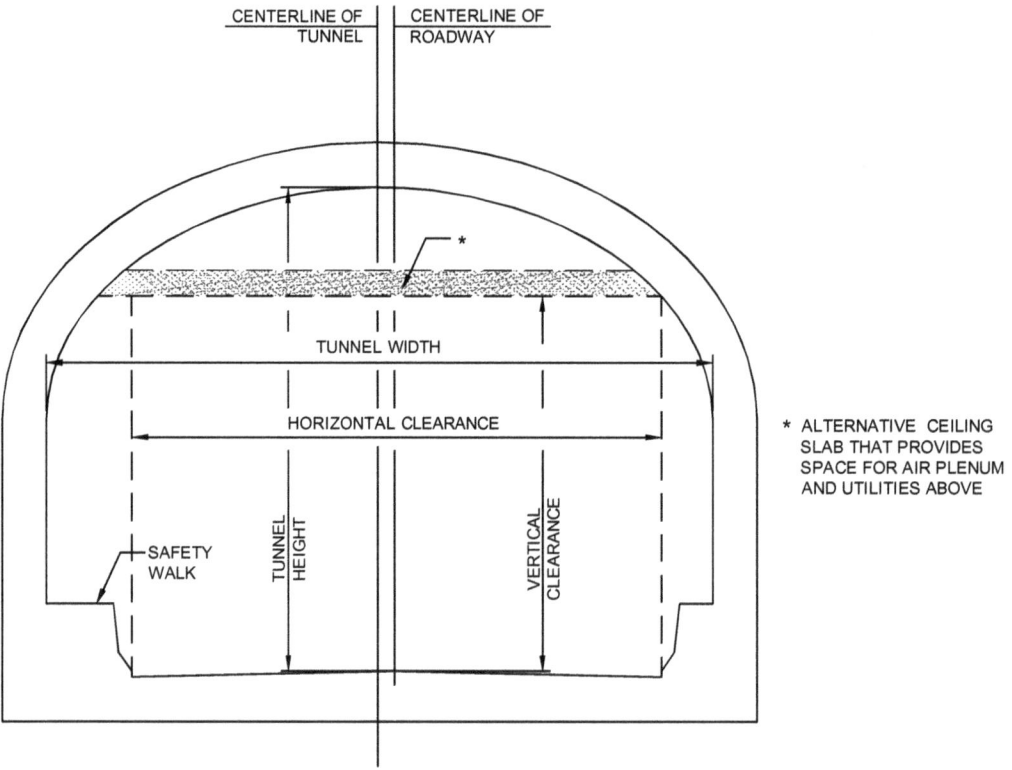

Figure 2.3 – Horseshoe tunnel with two traffic lanes and one safety walk. Also shown is an alternative ceiling slab. Invert may be a slab on grade or a structural slab.

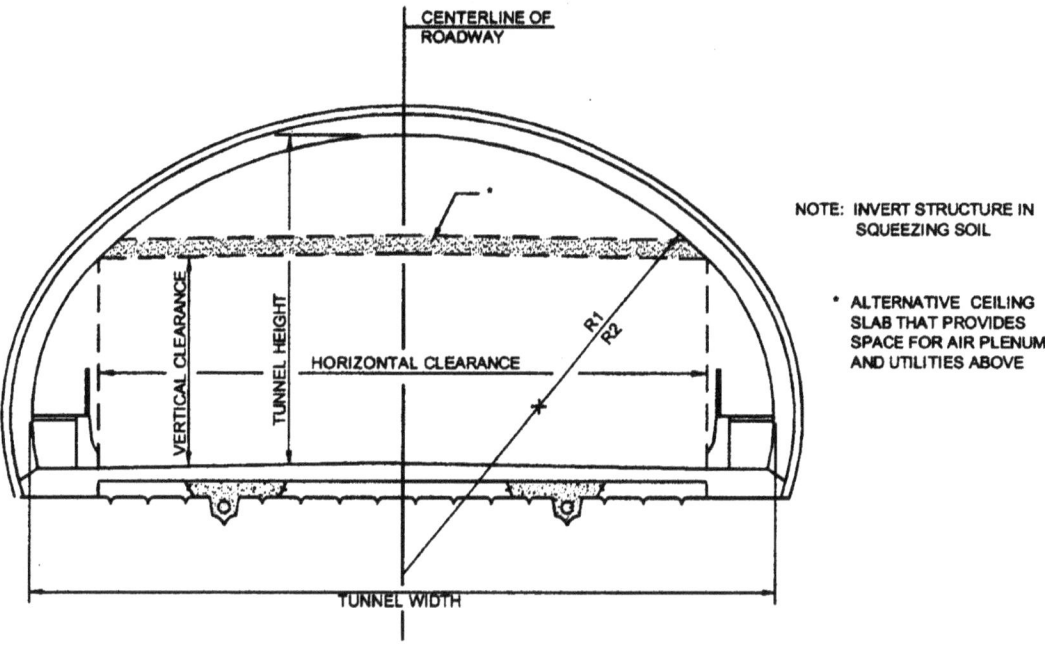

Figure 2.4 – Oval/egg tunnel with three traffic lanes and two safety walks. Also shown is alternative ceiling slab.

b) Rail Transit Tunnels

Figures 2.5 to 2.9 show the typical shapes for rail transit tunnels. As with highway tunnels, the shape typically relates to the method/ground conditions in which they were constructed. The shape of rail transit tunnels often varies along a given rail line. These shapes typically change at the transition between the station structure and the typical tunnel cross-section. However, the change in shape may also occur between stations due to variations in ground conditions.

Figure 2.5 – Circular tunnel with a single track and one safety walk. Invert slab is placed on top of liner.

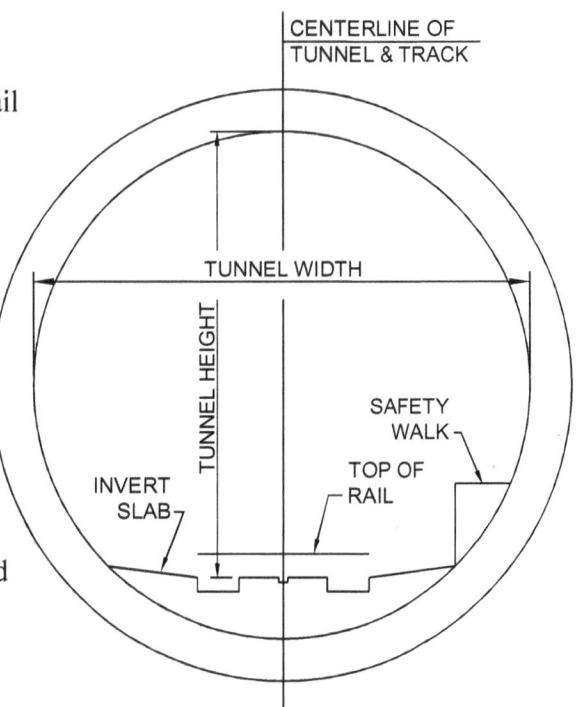

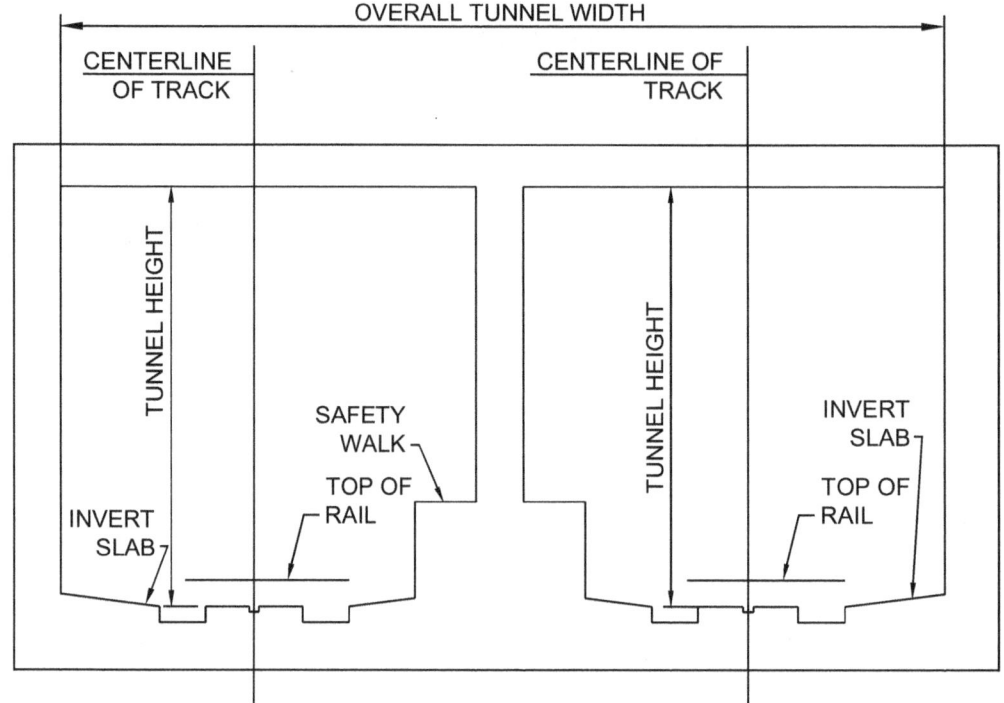

Figure 2.6 – Double box tunnel with a single track and one safety walk in each box. Depending on location and loading conditions, center wall may be solid or composed of consecutive columns.

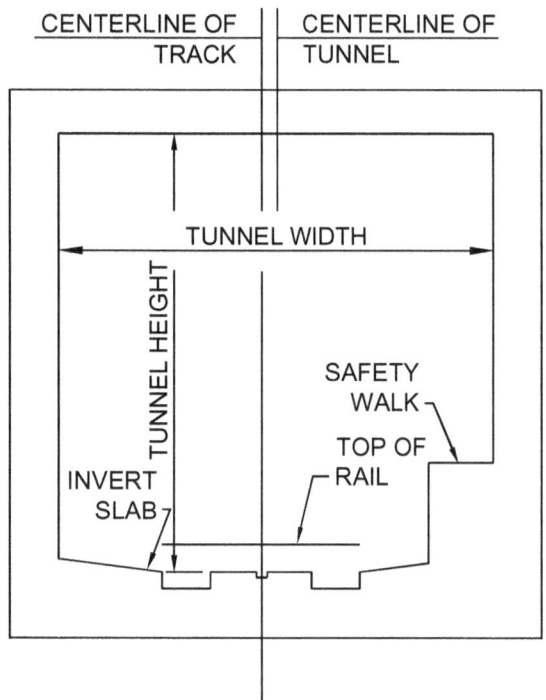

Figure 2.7 – Single box tunnel with a single track and one safety walk. Tunnel is usually constructed beside another single box tunnel for opposite direction travel.

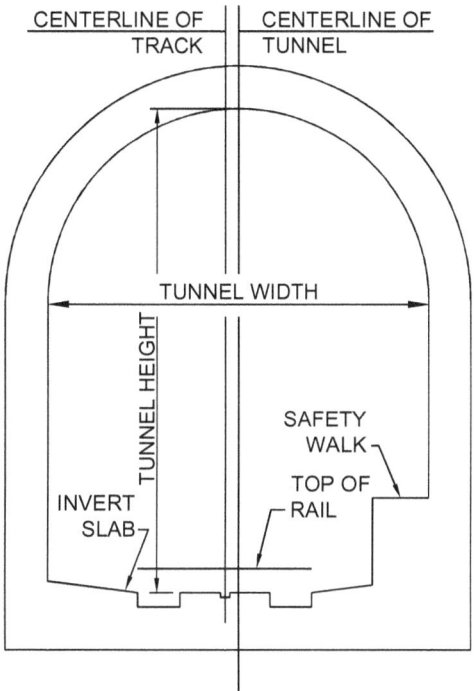

Figure 2.8 – Horseshoe tunnel with a single track and one safety walk. This shape typically exists in rock conditions and may be unlined within stable rock formations.

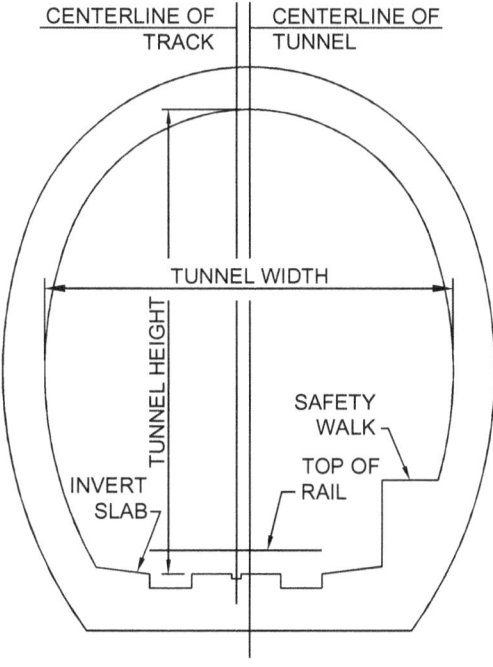

Figure 2.9 – Oval tunnel with a single track and one safety walk.

2. Liner Types

Tunnel liner types can be described using the following classifications:

- Unlined Rock
- Rock Reinforcement Systems
- Shotcrete
- Ribbed Systems
- Segmental Linings
- Poured Concrete
- Slurry Walls.

a) Unlined Rock

As the name suggests, an unlined rock tunnel is one in which no lining exists for the majority of the tunnel length. Linings of other types may exist at portals or at limited zones of weak rock. This type of liner was common in older railroad tunnels in the western mountains, some of which have been converted into highway tunnels for local access.

b) Rock Reinforcement Systems

Rock reinforcement systems are used to add additional stability to rock tunnels in which structural defects exist in the rock. The intent of these systems is to unify the rock pieces to produce a composite resistance to the outside forces. Reinforcement systems include the use of metal straps and mine ties with short bolts, untensioned steel dowels, or tensioned steel bolts. To prevent small fragments of rock from spalling off the lining, wire mesh, shotcrete, or a thin concrete lining may be used in conjunction with the above systems.

c) Shotcrete

Shotcrete is appealing as a lining type due to its ease of application and short "stand-up" time. Shotcrete is primarily used as a temporary application prior to a final liner being installed or as a local solution to instabilities in a rock tunnel. However, shotcrete can be used as a final lining. When this is the case, it is typically placed in layers and can have metal or randomly-oriented, synthetic fibers as reinforcement. The inside surface can be finished smooth as with regular concrete; therefore, it is difficult to determine the lining type without having knowledge of the construction method.

d) Ribbed Systems

Ribbed systems are typically a two-pass system for lining a drill-and-blast rock tunnel. The first pass consists of timber, steel, or precast concrete ribs usually with blocking between them. This provides structural stability to the tunnel. The second pass typically consists of poured concrete that is placed inside of the ribs. Another application of this system is to form the ribs using prefabricated reinforcing bar cages embedded in multiple layers of shotcrete. One other soft ground application is to place "barrel stave" timber lagging between the ribs.

e) Segmental Linings

Segmental linings are primarily used in conjunction with a tunnel boring machine (TBM) in soft ground conditions. The prefabricated lining segments are erected within the cylindrical tail shield of the TBM. These prefabricated segments can be made of steel, concrete, or cast iron and are usually bolted together to compress gaskets for preventing water penetration.

f) Placed Concrete

Placed concrete linings are usually the final linings that are installed over any of the previous initial stabilization methods. They can be used as a thin cover layer over the primary liner to provide a finished surface within the tunnel or to sandwich a waterproofing membrane. They can be reinforced or unreinforced. They can be designed as a non-structural finish element or as the main structural support for the tunnel.

g) Slurry Walls

Slurry wall construction types vary, but typically they consist of excavating a trench that matches the proposed wall profile. This trench is continually kept full with a drilling fluid during excavation, which stabilizes the sidewalls. Then a reinforcing cage is lowered into the slurry or soldier piles are driven at a predetermined interval and finally tremie concrete is placed into the excavation, which displaces the drilling fluid. This procedure is repeated in specified panel lengths, which are separated with watertight joints.

3. Invert Types

The invert of a tunnel is the slab on which the roadway or track bed is supported. There are two main methods for supporting the roadway or track bed; one is by placing the roadway or track bed directly on grade at the bottom of the tunnel structure, and the other is to span the roadway between sidewalls to provide space under the roadway for ventilation and utilities. The first method is used in most rail transit tunnels because their ventilation systems rarely use supply ductwork under the slab. This method is also employed in many highway tunnels over land where ventilation is supplied from above the roadway level.

The second method is commonly found in circular highway tunnels that must provide a horizontal roadway surface that is wide enough for at least two lanes of traffic and therefore the roadway slab is suspended off the tunnel bottom a particular distance. The void is then used for a ventilation plenum and other utilities. The roadway slab in many of the older highway tunnels in New York City is supported by placing structural steel beams, encased in concrete, that span transversely to the tunnel length, and are spaced between 750 mm (30 in) and 1,500 mm (60 in) on centers. Newer tunnels, similar to the second Hampton Roads Tunnel in Virginia, provide structural reinforced concrete slabs that span the required distance between supports.

It is necessary to determine the type of roadway slab used in a given tunnel because a more extensive inspection is required for a structural slab than for a slab-on-grade. Examples of structural slabs in common tunnel shapes are shown in Figures 2.10 to 2.12.

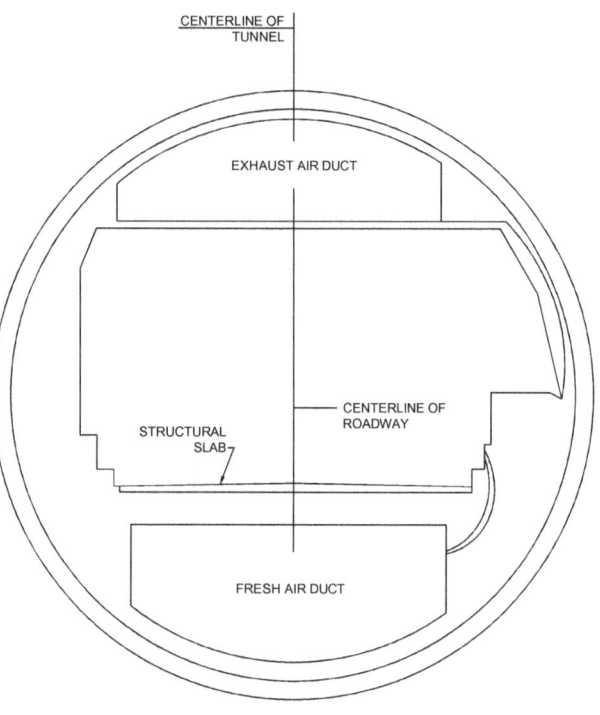

Figure 2.10 – Circular tunnel with a structural slab that provides space for an air plenum below.

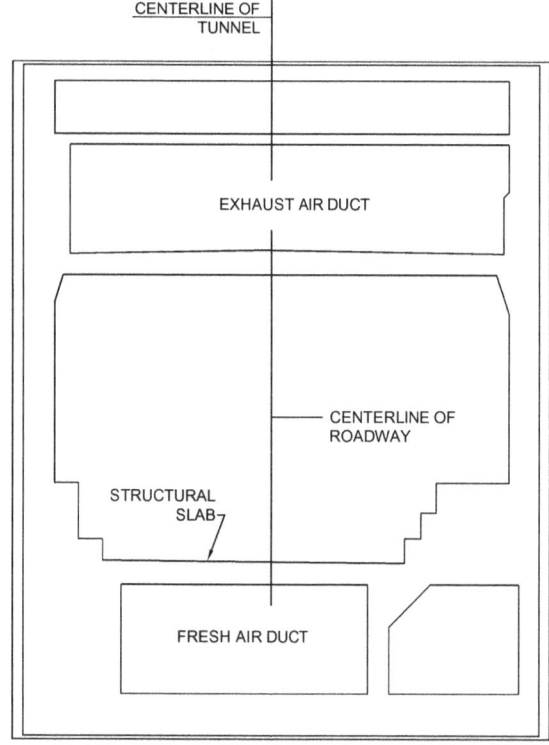

Figure 2.11 – Single box tunnel with a structural slab that provides space for an air plenum below.

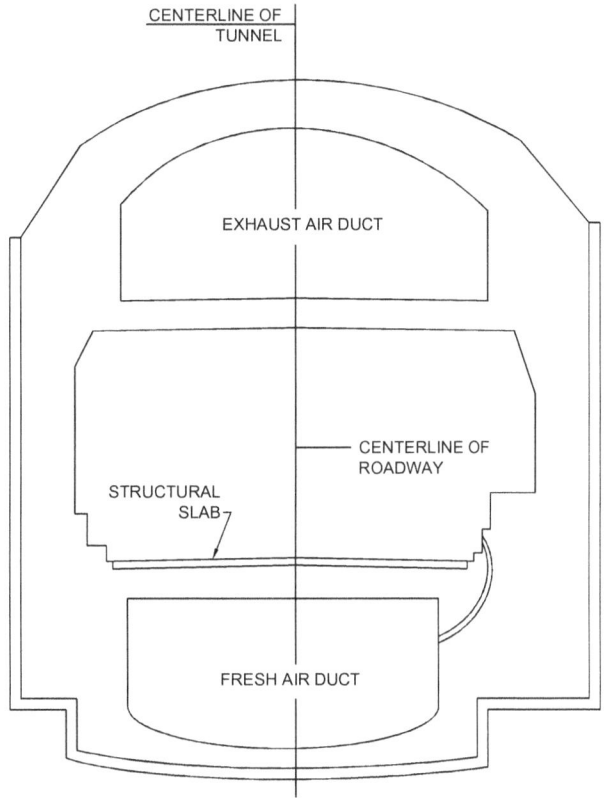

Figure 2.12 – Horseshoe tunnel with a structural slab that provides space for an air plenum below.

4. Construction Methods

As mentioned previously, the shape of the tunnel is largely dependent on the method used to construct the tunnel. Table 2.1 lists the six main methods used for tunnel construction with the shape that typically results. Brief descriptions of the construction methods follow:

	Circular	Horseshoe	Rectangular
Cut and Cover			X
Shield Driven	X		
Bored	X		
Drill and Blast	X	X	
Immersed Tube	X		X
Sequential Excavation		X	
Jacked Tunnels	X		X

Table 2.1 – Construction Methods

a) Cut and Cover

This method involves excavating an open trench in which the tunnel is constructed to the design finish elevation and subsequently covered with various compacted earthen materials and soils. Certain variations of this method include using piles and lagging, tie back anchors or slurry wall systems to construct the walls of a cut and cover tunnel.

b) Shield Driven

This method involves pushing a shield into the soft ground ahead. The material inside the shield is removed and a lining system is constructed before the shield is advanced further.

c) Bored

This method refers to using a mechanical TBM in which the full face of the tunnel cross section is excavated at one time using a variety of cutting tools that depend on ground conditions (soft ground or rock). The TBM is designed to support the adjacent soil until temporary (and subsequently permanent) linings are installed.

d) Drill and Blast

An alternative to using a TBM in rock situations would be to manually drill and blast the rock and remove it using conventional conveyor techniques. This method was commonly used for older tunnels and is still used when it is determined cost effective or in difficult ground conditions.

e) Immersed Tube

When a canal, channel, river, etc. needs to be crossed, this method is often used. A trench is dug at the water bottom and prefabricated tunnel segments are made water tight and sunken into position where they are connected to the other segments. Afterwards, the trench may be backfilled with earth to cover and protect the tunnel from the water borne traffic, e.g., ships, barges, and boats.

f) Sequential Excavation Method (SEM)

Soil in certain tunnels may have sufficient strength such that excavation of the soil face by equipment in small increments is possible without direct support. This excavation method is called the sequential excavation method. Once excavated, the soil face is then supported using shotcrete and the excavation is continued for the next segment. The cohesion of the rock or soil can be increased by injecting grouts into the ground prior to excavation of that segment.

g) Jacked Tunnels

The method of jacking a large tunnel underneath certain obstructions (highways, buildings, rail lines, etc.) that prohibit the use of typical cut-and-cover techniques for shallow tunnels has been used successfully in recent years. This method is considered when the obstruction cannot be moved or temporarily disturbed. First jacking pits are constructed. Then tunnel sections are constructed in the jacking pit and forced by large hydraulic jacks into the soft ground, which is systematically removed in front of the encroaching tunnel section. Sometimes if the soil above the proposed tunnel is poor then it is stabilized through various means such as grouting or freezing.

5. Tunnel Finishes

The interior finish of a tunnel is very important to the overall tunnel function. The finishes must meet the following standards to ensure tunnel safety and ease of maintenance:

- Be designed to enhance tunnel lighting and visibility
- Be fire resistant
- Be precluded from producing toxic fumes during a fire
- Be able to attenuate noise
- Be easy to clean.

A brief description of the typical types of tunnel finishes that exist in highway tunnels is given below. Transit tunnels often do not have an interior finish because the public is not exposed to the tunnel lining except as the tunnel approaches the stations or portals.

a) Ceramic Tile

This type of tunnel finish is the most widely used by tunnel owners. Tunnels with a concrete or shotcrete inner lining are conducive to tile placement because of their smooth surface. Ceramic tiles are extremely fire resistant, economical, easily cleaned, and good reflectors of light due to the smooth, glazed exterior finish. They are not, however, good sound attenuators, which in new tunnels has been addressed through other means. Typically, tiles are 106 mm (4 ¼ in) square and are available in a wide variety of colors. They differ from conventional ceramic tile in that they require a more secure connection to the tunnel lining to prevent the tiles from falling onto the roadway below. Even with a more secure connection, tiles may need to be replaced eventually because of normal deterioration. Additional tiles are typically purchased at the time of original construction since they are specifically made for that tunnel. The additional amount purchased can be up to 10 percent of the total tiled surface.

b) <u>Porcelain-Enameled Metal Panels</u>

Porcelain enamel is a combination of glass and inorganic color oxides that are fused to metal under extremely high temperatures. This method is used to coat most home appliances. The Porcelain Enamel Institute (PEI) has established guidelines for the performance of porcelain enamel through the following publications:

Appearance Properties (PEI 501)
Mechanical and Physical Properties (PEI 502)
Resistance to Corrosion (PEI 503)
High Temperature Properties (PEI 504)
Electrical Properties (PEI 505).

Porcelain enamel is typically applied to either cold-formed steel panels or extruded aluminum panels. For ceilings, the panels are often filled with a lightweight concrete; for walls, fiberglass boards are frequently used. The attributes of porcelain-enameled panels are similar to those for ceramic tile previously discussed; they are durable, easily washed, reflective, and come in a variety of colors. As with ceramic tile, these panels are not good for sound attenuation.

c) <u>Epoxy-Coated Concrete</u>

Epoxy coatings have been used on many tunnels during construction to reduce costs. Durable paints have also been used. The epoxy is a thermosetting resin that is chemically formulated for its toughness, strong adhesion, reflective ability, and low shrinkage. Experience has shown that these coatings do not withstand the harsh tunnel environmental conditions as well as the others, resulting in the need to repair or rehabilitate more often.

d) <u>Miscellaneous Finishes</u>

There are a variety of other finishes that can be used on the walls or ceilings of tunnels. Some of these finishes are becoming more popular due to their improved sound absorptive properties, ease of replacement, and ability to capitalize on the benefits of some of the materials mentioned above. Some of the systems are listed below:

(1) <u>Coated Cementboard Panels</u>

These panels are not in wide use in American tunnels at this time, but they offer a lightweight, fiber-reinforced cementboard that is coated with baked enamel.

(2) <u>Precast Concrete Panels</u>

This type of panel is often used as an alternative to metal panels; however, a combination of the two is also possible where the metal panel is applied as a veneer. Generally ceramic tile is cast into the underside of the panel as the final finish.

(3) <u>Metal Tiles</u>

This tile system is uncommon, but has been used successfully in certain tunnel applications. Metal tiles are coated with porcelain enamel and are set in mortar similarly to ceramic tile.

B. VENTILATION SYSTEMS

1. Types

Tunnel ventilation systems can be categorized into five main types or any combination of these five[2]. The five types are as follows:

- Natural Ventilation
- Longitudinal Ventilation
- Semi-Transverse Ventilation
- Full-Transverse Ventilation
- Single-Point Extraction.

It should be noted that ventilation systems are more applicable to highway tunnels due to high concentration of contaminants. Rail transit tunnels often have ventilation systems in the stations or at intermediate fan shafts, but during normal operations rely mainly on the piston effect of the train pushing air through the tunnel to remove stagnant air. Many rail transit tunnels have emergency mechanical ventilation that only works in the event of a fire. For further information on tunnel ventilation systems refer to NFPA 502 (National Fire Protection Association[10]).

a) <u>Natural Ventilation</u>

A naturally ventilated tunnel is as simple as the name implies. The movement of air is controlled by meteorological conditions and the piston effect created by moving traffic pushing the stale air through the tunnel. This effect is minimized when bi-directional traffic is present. The meteorological conditions include elevation and temperature differences between the two portals, and wind blowing into the tunnel. Figure 2.13 shows a typical profile of a naturally ventilated tunnel. Another configuration would be to add a center shaft that allows for one more portal by which air can enter or exit the tunnel. Many naturally ventilated tunnels over 180 m (600 ft) in length have mechanical fans installed for use during a fire emergency.

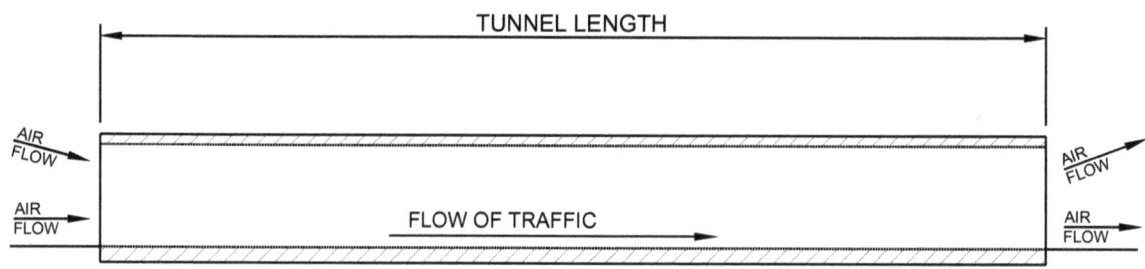

Figure 2.13 – Natural Ventilation

b)

Longitudinal Ventilation

Longitudinal ventilation is similar to natural ventilation with the addition of mechanical fans, either in the portal buildings, the center shaft, or mounted inside the tunnel. Longitudinal ventilation is often used inside rectangular-shaped tunnels that do not have the extra space

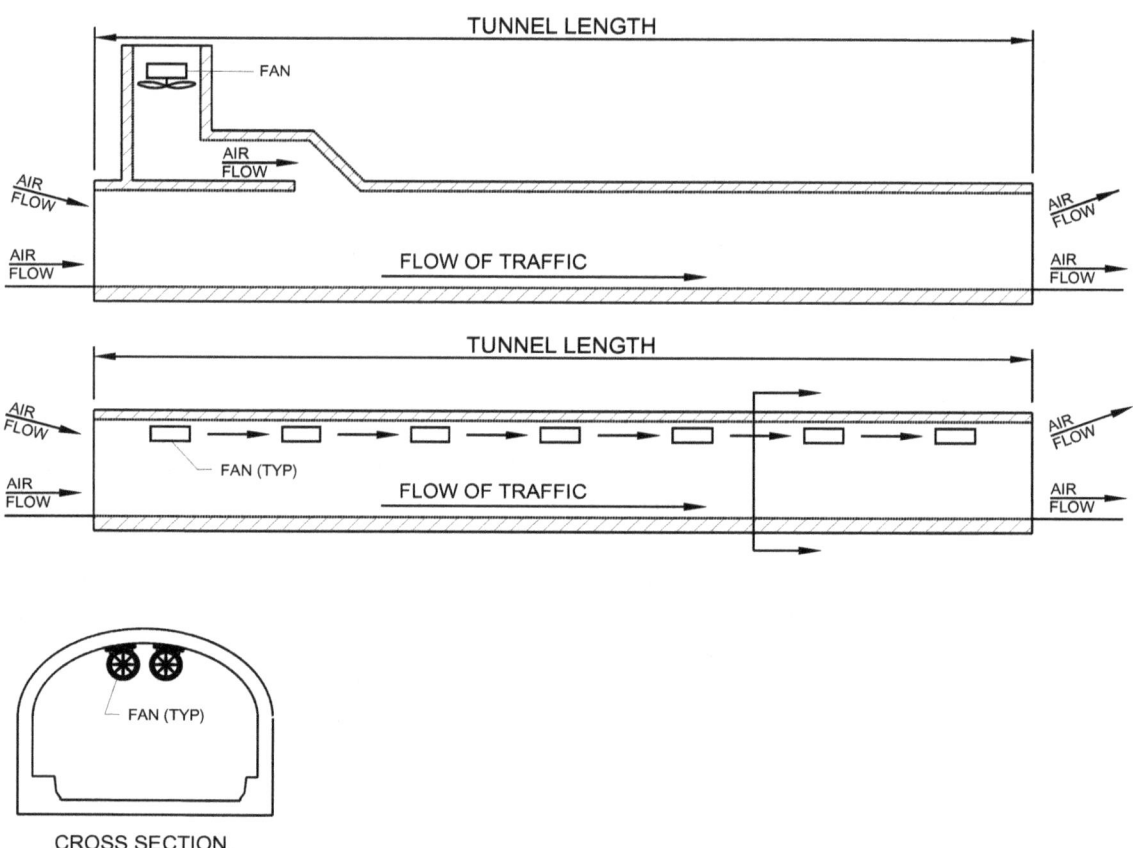

Figure 2.14 – Longitudinal Ventilation

above the ceiling or below the roadway for ductwork. Also, shorter circular tunnels may use the longitudinal system since there is less air to replace; therefore, the need for even distribution of air through ductwork is not necessary. The fans can be reversible and are used to move air into or out of the tunnel. Figure 2.14 shows two different configurations of longitudinally ventilated tunnels.

c) <u>Semi-Transverse Ventilation</u>

Semi-transverse ventilation also makes use of mechanical fans for movement of air, but it does not use the roadway envelope itself as the ductwork. A separate plenum or ductwork is added either above or below the tunnel with flues that allow for uniform distribution of air into or out of the tunnel. This plenum or ductwork is typically located above a suspended ceiling or below a structural slab within a tunnel with a circular cross-section. Figure 2.15 shows one example of a supply-air semi-transverse system and one example of an exhaust-air semi-transverse system. It should be noted that there are many variations of a semi-transverse system. One such variation would be to have half the tunnel be a supply-air system and the other half an exhaust-air system. Another variation is to have supply-air fans housed at both ends of the

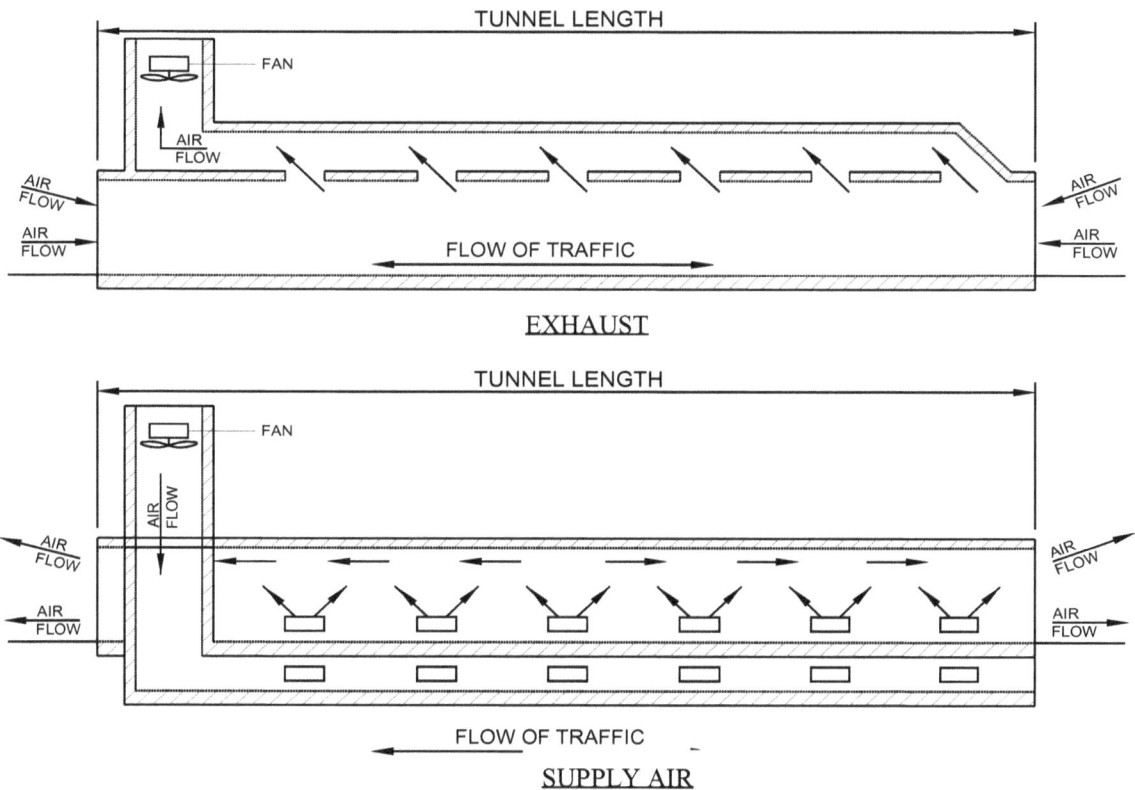

Figure 2.15 – Semi-Transverse Ventilation

plenum that push air directly into the plenum, towards the center of the tunnel. One last variation is to have a system that can either be exhaust-air or supply-air by utilizing reversible fans or a louver system in the ductwork that can change the direction of the air. In all cases, air either enters or leaves at both ends of the tunnel (bi-directional traffic flow) or on one end only (uni-directional traffic flow.)

d) <u>Full-Transverse Ventilation</u>

Full transverse ventilation uses the same components as semi-transverse ventilation, but it incorporates supply air and exhaust air together over the same length of tunnel. This method is used primarily for longer tunnels that have large amounts of air that need to be replaced or for heavily traveled tunnels that produce high levels of contaminants. The presence of supply and exhaust ducts allows for a pressure difference between the roadway and the ceiling; therefore, the air flows transverse to the tunnel length and is circulated more frequently. This system may also incorporate supply or exhaust ductwork along both sides of the tunnel instead of at the top and bottom. Figure 2.16 shows an example of a full-transverse ventilation system.

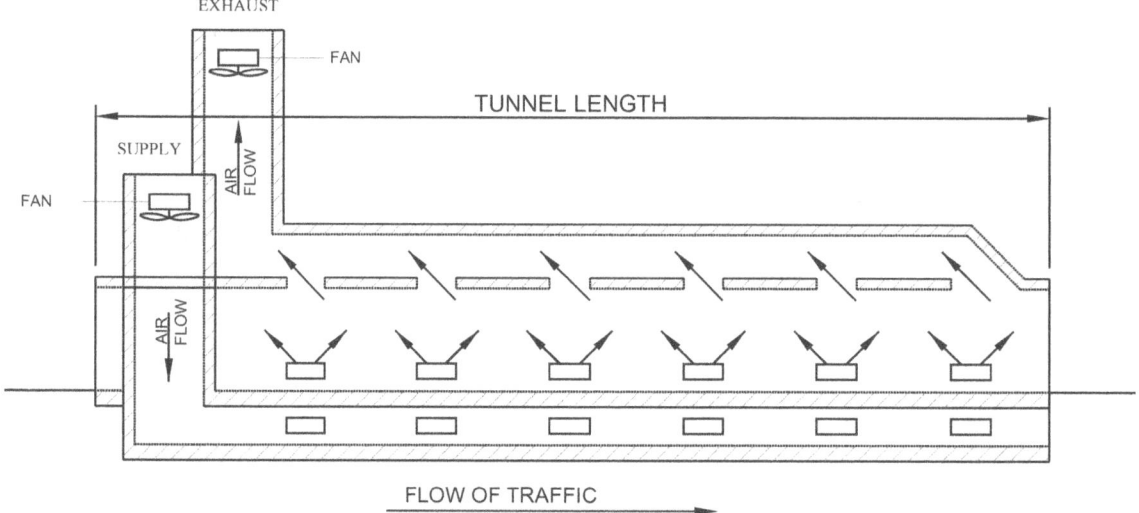

Figure 2.16 – Full-Transverse Ventilation

e) <u>Single-Point Extraction</u>

In conjunction with semi- and full-transverse ventilation systems, single-point extraction can be used to increase the airflow potential in the event of a fire in the tunnel. The system works by allowing the opening size of select exhaust flues to increase during an emergency. This can be done by mechanically opening louvers or by constructing portions of the ceiling out of material that would go from a solid to a gas during a fire, thus providing a larger opening. Both of these methods are rather costly and thus are seldom used. Newer tunnels achieve equal results simply by providing larger extraction ports at given intervals that are connected to the fans through the ductwork.

2. Equipment

a) Fans

(1) Axial

There are two main types of axial fans—tube axial fans and vane axial fans. Both types move air parallel to the impellor shaft, but the difference between the two is the addition of guide vanes on one or both sides of the impellor for the vane axial fans. These additional vanes allow the fan to deliver pressures that are approximately four times that of a typical tube axial fan. The two most common uses of axial fans are to mount them horizontally on the tunnel ceiling at given intervals along the tunnel or to mount them vertically within a ventilation shaft that exits to the surface.

Tube Axial Fan   Vane Axial Fan

Figure 2.17 – Axial Fans

(2) Centrifugal

This type of fan outlets the air in a direction that is 90 degrees to the direction at which air is obtained. Air enters parallel to the shaft of the blades and exits perpendicular to that. For tunnel applications, centrifugal fans can either be backward-curved or airfoil-bladed. Centrifugal fans are predominantly located within ventilation or portal buildings and are connected to supply or exhaust ductwork. They are commonly selected over axial fans due to their higher efficiency with less horsepower required and are therefore less expensive to operate.

Figure 2.18 – Centrifugal

b) Supplemental Equipment

   (1) Motors

Electric motors are typically used to drive the fans. They can be operated at either constant or variable speeds depending on the type of motor. According to the National Electric Manufacturers Association (NEMA), motors should be able to withstand a voltage and frequency adjustment of +/- 10 percent.

   (2) Fan Drives

A motor can be connected to the fan either directly or indirectly. Direct drives are where the fan is on the same shaft as the motor. Indirect drives allow for flexibility in motor location and are connected to the impellor shaft by belts, chains, or gears. The type of drive used can also induce speed variability for the ventilation system.

   (3) Sound Attenuators

Some tunnel exhaust systems are located in regions that require the noise generated by the fans to be reduced. This can be achieved by installing cylindrical or rectangular attenuators either mounted directly to the fan or within ductwork along the system.

   (4) Dampers

Objects used to control the flow of air within the ductwork are considered dampers. They are typically used in a full open or full closed position, but can also be operated at some position in between to regulate flow or pressure within the system.

C. LIGHTING SYSTEMS

1. Types

a) Highway Tunnels

There are various light sources that are used in tunnels to make up the tunnel lighting systems. These include fluorescent, high-pressure sodium, low-pressure sodium, metal halide, and pipe lighting, which is a system that may use one of the preceding light source types. Systems are chosen based on their life- cycle costs and the amount of light that is required for nighttime and daytime illumination. Shorter tunnels will require less daytime lighting due to the effect of light entering the portals on both ends, whereas longer tunnels will require extensive lighting for both nighttime and daytime conditions. In conjunction with the lighting system, a highly reflective surface on the walls and ceiling, such as tile or metal panels, may be used.

Fluorescent lights typically line the entire roadway tunnel length to provide the appropriate amount of light. At the ends of the roadway tunnel, low-pressure sodium lamps or high-pressure sodium lamps are often combined with the fluorescent lights to provide higher visibility when drivers' eyes are adjusting to the decrease in natural light. The transition length of tunnel required for having a higher lighting capacity varies from tunnel to tunnel and depends on which code the designer uses.

Both high-pressure sodium lamps and metal halide lamps are also typically used to line the entire length of roadway tunnels. In addition, pipe lighting, usually consisting of high-pressure sodium or metal halide lamps and longitudinal acrylic tubes on each side of the lamps, are used to disperse light uniformly along the tunnel length.

b) Rail Transit Tunnels

Rail transit tunnels are similar to highway tunnels in that they should provide sufficient light for train operators to properly adjust from the bright portal or station conditions to the darker conditions of the tunnel. Therefore, a certain length of brighter lights is necessary at the entrances to the tunnels. The individual tunnel owners usually stipulate the required level of lighting within the tunnel. However, as a minimum, light levels should be of such a magnitude that inspectors or workers at track level could clearly see the track elements without using flashlights.

D. OTHER SYSTEMS/APPURTENANCES

1. **Track**

The track system contains the following critical components:

a) Rail

The rail is a rolled, steel-shape portion of the track to be laid end-to-end in two parallel lines that the train or vehicle's wheels ride atop.

b) Rail Joints

Rail joints are mechanical fastenings designed to unite the abutting end of contiguous bolted rails.

c) Fasteners/Bolts/Spikes

These fasteners include a spike, bolt, or another mechanical device used to tie the rail to the crossties.

d) Tie Plates

Tie plates are rolled steel plates or a rubberized material designed to protect the timber crosstie from localized damage under the rails by distributing the wheel loads over a larger area. They assist in holding the rails to gage, tilt the rails inward to help counteract the outward thrust of wheel loads, and provide a more desirable positioning of the wheel bearing area on the rail head.

e) Crossties

Crossties are usually sawn solid timber, but may be made of precast reinforced concrete or fiber reinforced plastic. The many functions of a crosstie are to:

- Support vertical rail loads due to train weight.
- Distribute those loads over a wide area of supporting material.
- Hold fasteners that can resist rail rotation due to laterally imposed loads.
- Maintain a fixed distance between the two rails making up a track.
- Help keep the two rails at the correct relative elevation.
- Anchor the rails against both lateral and longitudinal movement by embedment in the ballast.
- Provide a convenient system for adjusting the vertical profile of the track.

f) Ballast

Ballast is a coarse granular material forming a bed for ties, usually rocks. The ballast is used to transmit and distribute the load of the track and railroad rolling equipment to the sub-grade; restrain the track laterally, longitudinally, and vertically under dynamic loads imposed by railroad rolling equipment and thermal stresses exerted by the rails; provide adequate drainage for the track; and maintain proper cross-level surface and alignment.

g) Plinth Pads

Plinth pads are concrete support pads or pedestals that are fastened directly to the concrete invert. These pads are placed at close intervals and permit the rail to span directly from one pad to another.

2. Power (Third Rail/Catenary)

a) Third Rail Power System

A third rail power system will consist of the elements listed below and will typically be arranged as shown in Figures 2.19 and 2.20

(1) Steel Contact Rail

Steel contact rail is the rail that carries power for electric rail cars through the tunnel and is placed parallel to the other two standard rails.

(2) Contact Rail Insulators

Contact rail insulators are made either of porcelain or fiberglass and are to be installed at each supporting bracket location.

(3) Protection Board

Protection boards are placed above the steel contact rail to "protect" personnel from making direct contact with this rail. These boards are typically made of fiberglass or timber.

(4) Protection Board Brackets

Protection board brackets are mounted on either timber ties or concrete ties/base and are used to support the protection board at a distance above the steel contact rail.

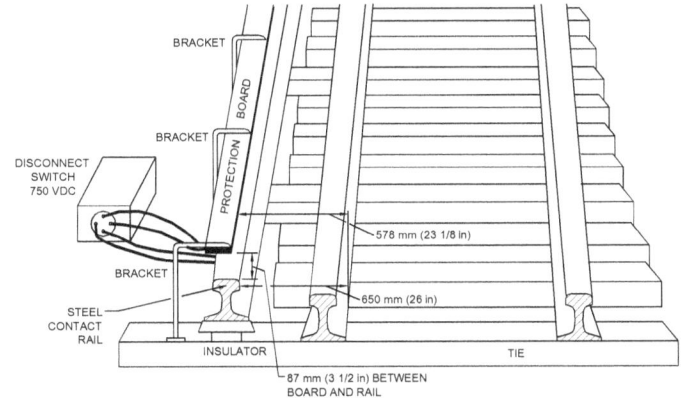

Figure 2.19 – Typical Third Rail Power System
(Note: Dimensions indicate minimum clearance requirements)

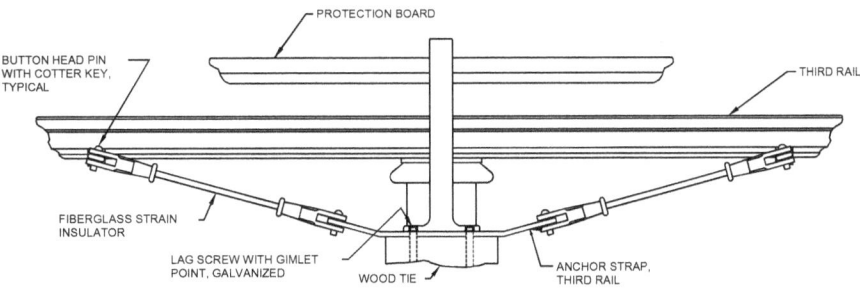

Figure 2.20 – Typical Third Rail Insulated Anchor Arm

(5)   Third Rail Insulated Anchor Arms

Third rail insulated anchor arms are located at the midpoint of each long section, with a maximum length for any section limited to 1.6 km (1 mile).

b)   Catenary Power System

The catenary system is an overhead power system whereby the rail transit cars are powered by means of contact between the pantographs atop the rail car and the catenary wire. A typical catenary system may consist of some or all of the following components: balance weights, yoke plates, steady arms, insulators, hangers, jumpers, safety assemblies, pull-off arrangements, back guys and anchors, underbridge assemblies, contact wires, clamped electrical connectors, messenger supports, registration assemblies, overlaps, section insulators, phase breaks, and section disconnects. For tunnel catenary systems, some of the above components are not necessary or are modified in their use. This is particularly true for the methods of support in that the catenary system is supported directly from the tunnel structure instead of from poles with guy wires.

Since the methods used to support a catenary system within a tunnel can vary, a detailed description of the individual components is not given in this section. For inspection purposes, Chapter 4, Section D, Part 2 provides inspection procedures for various components listed above that may exist in a tunnel catenary system.

3. Signal/Communication Systems

a)   Signal System

The signal system is a complex assortment of electrical and mechanical instruments that work together to provide direction for the individual trains within a transit system. A typical signal system may consist of some or all of the following components: signals, signal cases, relay rooms, switch machines, switch circuit controllers, local cables, express cables, signal power cables, duct banks, messenger systems, pull boxes, cable vaults, transformers, disconnects, and local control facilities.

b)   Communication System

The communication system consists of all devices that allow communication from or within a tunnel. Examples of these systems would be emergency phones that are located periodically along a highway tunnel and radios by which train controllers correspond with each other and central operations. The specific components included in a communication system include the phones and radios, as well as any cables, wires, or other equipment that is needed to transport the messages.

CHAPTER 3:
FUNDAMENTALS OF TUNNEL INSPECTION

A. INSPECTOR QUALIFICATIONS

The inspection should be accomplished with teams consisting of a minimum of two individuals; all team members should meet certain minimum qualifications as defined below. Aside from these general qualifications and the specific qualifications listed for each discipline, a tunnel owner may require that all tunnel inspection team members be certified to ensure that these qualifications are met. This certification would need to be performed by the tunnel owner, since currently there is no national tunnel inspection certification program. These inspection team members are classified as a Team Leader and Team Member(s). All individuals who will perform inspection work should be knowledgeable of tunnel components and understand how they function. The inspection team should meet the following general qualifications:

- Be able to climb and/or use equipment to access the higher regions of the structures.
- Be able to evaluate and determine types of equipment or testing required to fully define a deficiency.
- Be able to print legibly and to draw understandable sketches.
- Be able to read and interpret drawings.
- Be able to use tablet PC's for data collection.

Specific qualifications of the Team Leader and Team Member(s) for the various components to be inspected include:

1. Civil/Structural

a) Team Leader

- Be a registered professional engineer or
- Have design experience in tunnels using the same materials and
- Have a minimum of five years inspection experience with the ability to identify and evaluate defects that pose a threat to the integrity of a structural member.
- Be able to assess the degree of deterioration for concrete, steel, masonry, and timber members.

b) Team Member(s)

- Be trained in general tunnel inspection requirements.
- Have a minimum of one year inspection experience in concrete, steel, timber, and masonry structures.

2. **Mechanical**

   a) Team Leader

   - Be a registered professional engineer or
   - Have design experience or be familiar with the type of mechanical systems installed in the tunnel. Examples of these systems are, but not limited to:

     - Tunnel Ventilation
     - Air Conditioning
     - Heating
     - Controls
     - Plumbing
     - Tunnel Drainage Systems (e.g., sump pumps)
     - Fire Protection
     - Wells/Septic.

   - Have a minimum of three years inspection experience with the ability to evaluate the physical condition as well as the operational condition of equipment.
   - Be aware of applicable codes and guidelines for tunnel construction and operation pertaining to mechanical features.

   b) Team Member(s)

   - Be trained in general inspection requirements.
   - Have a minimum of one year inspection experience with mechanical and plumbing systems.

3. **Electrical**

   a) Team Leader

   - Be a registered professional engineer or
   - Have design experience or be familiar with the type of electrical systems installed in the tunnel. Examples of these systems include, but are not limited to:

     - Power Distribution
     - Emergency Power
     - Lighting
     - Fire Detection
     - Communications.

- Have a minimum of three years inspection experience with the ability to evaluate the physical condition as well as the operational condition of the electrical systems and equipment.

- Be aware of applicable codes and guidelines for tunnel construction and operation, including, but not limited to the following:

    - NETA MTS 1 – *National Electrical Testing Association, Maintenance Testing Specifications* – developed for those responsible for the continued operation of existing electrical systems and equipment to guide them in specifying and performing the necessary tests to ensure that these systems and apparatus perform satisfactorily, minimizing downtime and maximizing life expectancy.
    - NFPA 70 – *National Fire Protection Association 70* – covers installations of electric conductors and equipment within or on public and private buildings or other structures, installations of conductors and equipment that connect to the supply of electricity, installations of other outside conductors and equipment on the premises, and installations of optical fiber cables and raceways.
    - NFPA 70B – *National Fire Protection Association 70B* – recommended practice for electrical equipment maintenance for industrial-type electrical systems and equipment, but is not intended to duplicate or supersede instructions that electrical manufacturers normally provide.
    - NFPA 70E – *National Fire Protection Association 70E* – addresses those electrical safety requirements for employee workplaces that are necessary for the practical safeguarding of employees in their pursuit of gainful employment.
    - NFPA 72 – *National Fire Protection Association 72* – national fire alarm code that covers the application, installation, location, performance, and maintenance of fire alarm systems and their components.
    - NFPA 130 – *National Fire Protection Association 130* – covers fire protection requirements for passenger rail, underground, surface, and elevated fixed guideway transit systems including trainways, vehicles, fixed guideway transit stations, and vehicle maintenance and storage areas; and for life safety from fire in fixed guideway transit stations, trainways, vehicles, and outdoor vehicle maintenance and storage areas.
    - IES LM-50 – *Illuminating Engineering Society, Lighting Measurements–50* – provides a uniform test procedure for determining, measuring, and reporting the illuminance and luminance characteristics of roadway lighting installations.

- IES RP-22 – *Illuminating Engineering Society, Recommended Practices–22* – provides information to assist engineers and designers in determining lighting needs, recommending solutions, and evaluating resulting visibility at vehicular tunnel approaches and interiors.

b) <u>Team Member(s)</u>

- Be trained in general electrical inspection requirements.
- Have a minimum of one year inspection experience with electrical systems.

c) <u>Special Testing Agencies</u>

The use of special testing agencies is required for testing the power distribution and fire protection systems. Such agencies shall meet the following requirements:

- Be a member of The International Electrical Testing Association (NETA) or meet all of the following qualifications:

  - Be nationally recognized as an electrical testing laboratory.
  - Be regularly engaged in the testing of electrical systems and equipment for the past five years.
  - Have at least one professional engineer on staff that is licensed in the state where the work is being done.
  - Have in house or lease sufficient calibrated equipment to do the testing required.
  - Have a means to trace all test instrument calibration to The National Bureau of Standards.

4. **Track, Third Rail, Catenary, Signals, and Communications**

   a) <u>Team Leader</u>

   - Be a registered professional engineer or
   - Have design experience in the system components being inspected.
   - Have a minimum of three years inspection experience with the ability to identify and evaluate the condition of the components being inspected.
   - Be familiar with specialized testing equipment to assess a particular component's operational viability.

b) <u>Team Member(s)</u>

- Be trained in general inspection requirements.
- Have a minimum of one year inspection experience with track, third rail, catenary, signals and communications systems.

B. **RESPONSIBILITIES**

Each member of the inspection team must fulfill certain duties for work to be accomplished in an efficient manner. The Team Leader is responsible for coordination with appropriate tunnel and supervisory staff for access into the tunnels, for scheduling equipment, for determining the degree of inspection required, for evaluating all deficiencies, for ensuring that all inspection forms are thoroughly completed and legible (if using paper forms), and for notifying appropriate tunnel staff of any potentially dangerous condition. The other Team Member will assist the team leader in the inspection. Such duties may include performing portions of the inspection, carrying the equipment and inspection forms, taking photographs, and making sketches.

The tunnel owner is responsible for closing down the tunnel for inspection access and for responding to any critical actions that are identified by the inspectors.

C. **EQUIPMENT/TOOLS**

Below is a suggested list of equipment and tools commonly used for tunnel inspections:

- Aerial Bucket Truck or High Lift - Used to lift the inspector to areas inaccessible by foot or ladders.
- Hi-Rail Vehicle - Used atop the track to gain access to the rail transit tunnel structure or catenary power system. The tunnel owner may supply such vehicles.
- Awl/Boring Tool - Used to determine extent of deterioration in timber.
- Calipers - Used to measure steel plate thicknesses.
- Camera (35mm or digital) with Flash - Used to take photographs for documentation of the inspection.
- Chalk, Keel, or Markers - Used to make reference marks on tunnel surfaces.
- Chipping Hammer - Used to sound concrete (see Chapter 4, Section A, Part 2 – What to Look For).
- Clip Board - Used to take notes and fill out paper inspection forms during the inspection.
- Crack Comparator Gauge - Used to measure crack widths in inches.
- D – Meter - Used to measure the thickness of steel.
- Extension Cord - Used to get electricity to inspection area.
- Field Forms - Used to document the findings, take notes, and draw sketches for the various structures.
- Flashlights - Used in dark areas to help see during inspection.
- Ladders - Used in lieu of a lifting system to access areas not visible from the ground.

- Light Meter - Used to measure the brightness in the tunnel.
- Halogen Lights - Used where tunnel lighting is inadequate during inspection.
- Pencil
- Plumb Bob - Used to check plumbness of columns and wall faces.
- Pocket Knife - Used to examine loose material and other items.
- Sample Bottles - Used to obtain liquid samples.
- Scraper - Used to determine extent of corrosion and concrete deterioration.
- Screw Driver - Used to probe weep holes to check for clogs.
- Wire Brush or Brooms - Used to clean debris from surfaces to be inspected.
- Tablet PC - Used to take notes or draw sketches onto screens that would be synonymous with paper forms.
- Tapes
  - Pocket Tapes and Folding Rules - Used to measure dimensions of defects.
  - 30 m (100 ft) Tape (Non Metallic) - Used to measure anything beyond the reach of pocket tapes and folding rules.

Safety equipment meeting the most current OSHA Standards should be available for the inspection team's use and may include:

- Appropriate devices for traffic control
- First aid kit
- Flashlights
- Hardhats
- Leather work gloves
- Appropriate safety vests
- Protective eye wear
- Knee pads
- Safety belts or harnesses
- Work boots
- Protective breathing masks if soot and dirt buildup is prevalent on the tunnel surfaces
- Air quality monitoring equipment.

For the confined spaces, the appropriate equipment designated by the field safety officer should be employed. This equipment includes respirators, tie off ropes, radios, and instruments to measure gas levels. It is especially important to monitor gas levels in areas of known ground contamination by deleterious materials.

D. **PREPARATION**

1. **Mobilization**

Prior to conducting tunnel inspections, a mobilization period of planning and organizing for the inspection is a must for the inspection to be performed as efficiently as possible.

If the inspection is to be conducted by a consultant, the consultant will need to coordinate carefully with the tunnel owner to determine available access times for inspecting within the road or track area, where vehicles can be parked, communication procedures for shutting off and locking out fans during the inspection, timing for shutting down electrical systems for testing, discussion of known problem areas, etc.

A vital part of the mobilization phase is the receipt and study of available tunnel drawings or previous inspection reports. It is crucial to minimize tunnel closures; therefore, forms should be developed on paper or on computer screens during the mobilization period prior to entering the field. The forms should also contain the necessary fields of information to be supplied as part of the computerized database. It is also critical that all health and safety plans, where confined space entry is deemed necessary, be completed and inspectors be knowledgeable of their responsibilities.

In summary, the planning and scheduling of the inspection during the mobilization phase should lead to an efficiently run inspection effort that benefits both the inspection team and the tunnel owner/operator.

2. **Survey Control**

It is necessary to establish a system by which the location of a defect can be recorded and understood in reference to where the defect is observed. Establishing such a system will allow the inspections to be referenced historically for future monitoring of the condition of a particular defect and will increase the efficiency of the overall inspection process.

Most highway and rail transit tunnels already have a baseline or stationing system established throughout the tunnel. This allows defects to be recorded using the station where they occur. Some tunnel owners have defined panels that are of a given length and sequentially numbered. Joints in the lining material are used to delineate these panels. To tie the panels into the baseline system, the station of the beginning and end of each panel can be established and a defect can be located relative to its distance from either end of the panel, which can subsequently be converted into a specific station or distance from the end of the tunnel.

In addition to locating a defect by panel number and station, it is necessary to note the defect's position within the tunnel cross-section. Figures 3.1 to 3.5 show a typical tunnel layout plan along with the designations for typical tunnel cross-sections. Defects in circular tunnels without air ducts or structural slabs can be located using a clock system with 12:00 being at the top. Horseshoe, rectangular, and other circular tunnels can be broken down into the cross-sectional elements that are shown on the following pages.

Table 3.1- Tunnel Inspection Layout Plan

TUNNEL INSPECTION LAYOUT PLAN

★ DENOTES PANEL NUMBER IF AVAILABLE

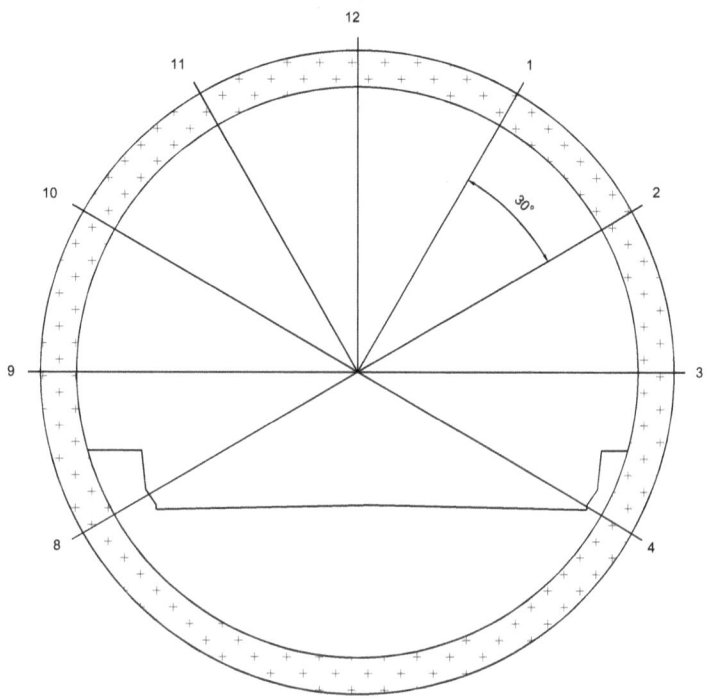

Figure 3.2 – Circular Tunnel Clock System Designations

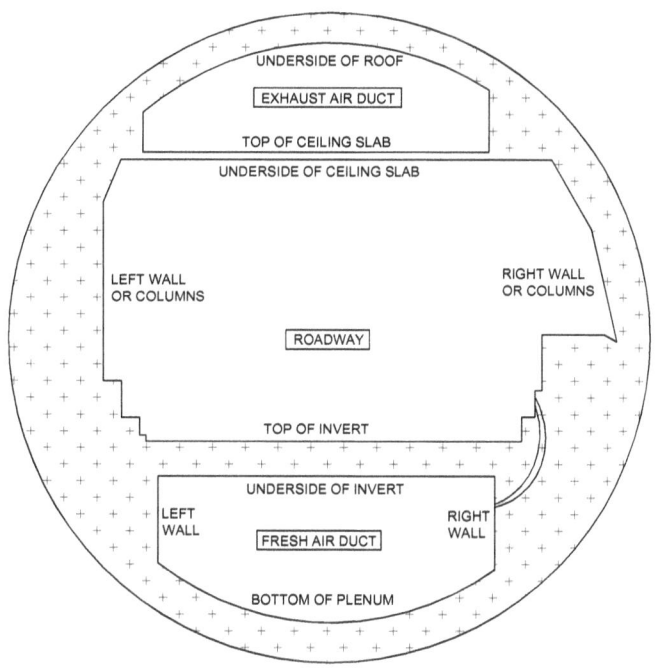

Figure 3.3 – Circular Tunnel Label System Designations

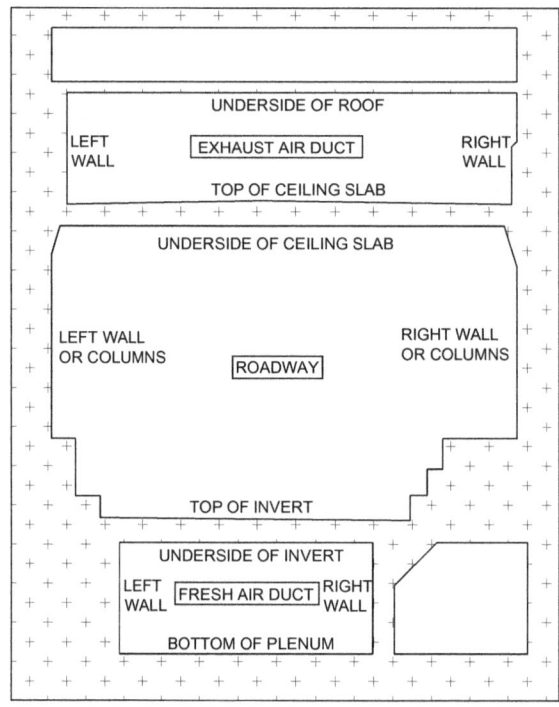

Figure 3.4 – Rectangular Tunnel Label System Designations

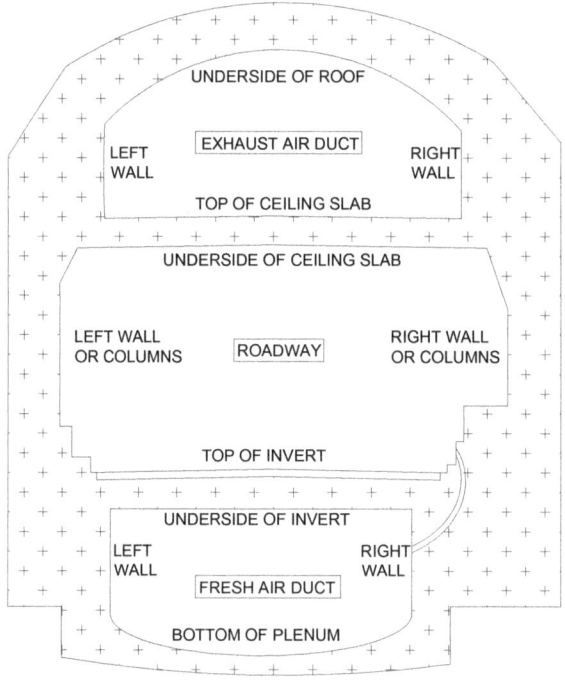

Figure 3.5 – Horseshoe Tunnel Label System Designations

3. Inspection Forms

To properly gather and record structural inspection data for historical purposes, it is necessary to develop forms that are clearly understood and easily entered into a database. These forms can be on pre-printed sheets and then scanned into the database or transferred manually. Another option would be to make use of tablet PC's, which can be pre-programmed to correspond to the main database. This option allows the data to be downloaded directly into the structural database for future use.

The forms to be used fall into two main categories: documentation forms and defect location forms. The documentation forms are discussed in this section along with examples of forms to be used for both highway and rail transit tunnels. Examples of the defect location forms that apply to the components of main tunnel segments or panels as well as auxiliary spaces such as portal buildings, control rooms, stations, etc., are provided in Chapter 5, Section A.

Below are general instructions for proper completion of the subsequent example documentation forms:

a) Highway Forms

(1) Condition Code Form

Tunnel Name: Enter the name typically assigned to the tunnel.

Begin Station: Enter the beginning station of the tunnel segment for which this form is being completed (e.g., Sta. 70+00.00).

End Station: Enter the ending station of the tunnel segment for which this form is being completed (e.g., Sta. 72+00.00).

Panel Number: If present, enter the predetermined panel number for the segment for which this form is being completed (e.g., 101).

Year Built: Enter the year during which construction of the tunnel was completed.

Liner Type: Enter the appropriate acronym from Table 3.1.

Date of Inspection: Enter the month, day, and year the inspection is performed.

Table 3.1 – Liner Type Acronyms

Acronym	Description
UR	Unlined Rock
CIPNR	Cast-In-Place Concrete, No Reinforcement
CIPR	Cast-In-Place Concrete, Reinforced
SG	Shotcrete/Gunite
PCLS	Precast Concrete Liner Segments
SILP	Steel/Iron Liner Plate
M	Masonry
T	Timber
SCB	Steel Columns and Beams, Jackarches
RMPS	Rock-fall Mesh Pinned to Surface

Inspector(s): Enter the inspector(s) first initial and last name.

Condition Codes: Assign each element a numerical rating in accordance with the rating schedules given in Chapter 4, Section A, Part 4. Enter a dash (-) for all elements that are not present in the tunnel segment.

Comments: Add any pertinent comments as necessary for properly explaining the tunnel segment's condition codes.

(2) <u>Supplemental Tunnel Segment Sketches</u>

Tunnel Name: Enter the name typically assigned to the tunnel.

Begin Station: Enter the beginning station of the tunnel segment for which this form is being completed (e.g., Sta. 70+00.00).

End Station: Enter the ending station of the tunnel segment for which this form is being completed (e.g., Sta. 72+00.00).

Panel Number: If present, enter the predetermined panel number for the segment for which this form is being completed (e.g. 101).

Year Built: Enter the year during which construction of the tunnel was completed.

Liner Type: Enter the appropriate acronym from Table 3.1.

Date of Inspection: Enter the month, day, and year the inspection is performed.

Inspector(s): Enter the inspector(s) first initial and last name.

Sketches: Provide detailed sketches of defects found in areas of the tunnel or auxiliary spaces that is not covered by the standard forms that are provided in Chapter 5, Section A.

(3) <u>Tunnel Segment Photo Log Sheet</u>

Tunnel Name: Enter the name typically assigned to the tunnel.

Begin Station: Enter the beginning station of the tunnel segment for which this form is being completed (e.g., Sta. 70+00.00).

End Station: Enter the ending station of the tunnel segment for which this form is being completed (e.g., Sta. 72+00.00).

Panel Number: If present, enter the predetermined panel number for the segment for which this form is being completed (e.g., 101).

Year Built: Enter the year during which construction of the tunnel was completed.

Liner Type: Enter the appropriate acronym from Table 3.1.

Date of Inspection: Enter the month, day, and year the inspection is performed.

Inspector(s): Enter the inspector(s) first initial and last name.

Photos: Provide relevant information for any photos that are taken within that tunnel segment. Include as much detail as possible within the description section. It would be helpful to take photographs of the same conditions or defects as previous inspections, so that the rate of deterioration can be ascertained.

b) Rail Transit Forms

(1) Condition Code Form

Line: Enter the given name of the system line that is being inspected (e.g., A, Blue, Market Street, etc.).

Track: Enter directional description of track being inspected (e.g., inbound, outbound, north, east, south, west, etc.).

Name of Station Ahead: Enter name of station (subway or ground) that is next along direction of train traffic.

Name of Station Behind: Enter name of station (subway or ground) that is next in opposite direction of train traffic.

Begin Station: Enter the beginning station of the tunnel segment for which this form is being completed (e.g., Sta. 70+00.00).

End Station: Enter the ending station of the tunnel segment for which this form is being completed (e.g., Sta. 72+00.00).

Panel Number: If present, enter the predetermined panel number for the segment for which this form is being completed (e.g., 101).

Year Built: Enter the year during which construction of the tunnel was completed.

Liner Type: Enter the appropriate acronym from Table 3.1.

Date of Inspection: Enter the month, day, and year the inspection is performed.

Inspector(s): Enter the inspector(s) first initial and last name.

Condition Codes: Assign each element a numerical rating in accordance with the rating schedules given in Chapter 4, Section A, Part 4. Enter a dash (-) for all elements that are not present in the tunnel segment.

Comments: Add any pertinent comments as necessary for properly explaining the tunnel segment's condition codes.

(2) <u>Supplemental Tunnel Segment Sketches</u>

Line: Enter the given name of the system line that is being inspected (e.g., A, Blue, Market Street, etc.).

Track: Enter directional description of track being inspected (e.g., inbound, outbound, north, east, south, west, etc.).

Name of Station Ahead: Enter name of station (subway or ground) that is next along direction of train traffic.

Name of Station Behind: Enter name of station (subway or ground) that is next in opposite direction of train traffic.

Begin Station: Enter the beginning station of the tunnel segment for which this form is being completed (e.g., Sta. 70+00.00).

End Station: Enter the ending station of the tunnel segment for which this form is being completed (e.g., Sta. 72+00.00).

Panel Number: If present, enter the predetermined panel number for the segment for which this form is being completed (e.g., 101).

Year Built: Enter the year during which construction of the tunnel was completed.

Liner Type: Enter the appropriate acronym from Table 3.1.

Date of Inspection: Enter the month, day, and year the inspection is performed.

Inspector(s): Enter the inspector(s) first initial and last name.

Sketches: Provide detailed sketches of defects found in areas of the tunnel or auxiliary spaces that is not covered by the standard forms that are provided in Chapter 5, Section A.

(3) <u>Tunnel Segment Photo Log Sheet</u>

Line: Enter the given name of the system line that is being inspected (e.g., A, Blue, Market Street, etc.).

Track: Enter directional description of track being inspected (e.g., inbound, outbound, north, east, south, west, etc.).

Name of Station Ahead: Enter name of station (subway or ground) that is next along direction of train traffic.

Name of Station Behind: Enter name of station (subway or ground) that is next in opposite direction of train traffic.

Begin Station: Enter the beginning station of the tunnel segment for which this form is being completed (e.g., Sta. 70+00.00).

End Station: Enter the ending station of the tunnel segment for which this form is being completed (e.g., Sta. 72+00.00).

Panel Number: If present enter the predetermined panel number for the segment for which this form is being completed (e.g., 101).

Year Built: Enter the year during which construction of the tunnel was completed.

Liner Type: Enter the appropriate acronym from Table 3.1.

Date of Inspection: Enter the month, day, and year the inspection is performed.

Inspector(s): Enter the inspector(s) first initial and last name.

Photos: Provide relevant information for any photos that are taken within that tunnel segment. Include as much detail as possible within the description section. It would be helpful to take photographs of the same conditions or defects as previous inspections, so that the rate of deterioration can be ascertained.

TUNNEL AGENCY NAME
HIGHWAY TUNNEL FIELD INSPECTION FORM
TUNNEL SEGMENT CONDITION CODES

General Information

Tunnel Name _____

Begin Station _____ End Station _____

Or

Panel Number _____

Year Built _____   Liner Type _____

Date of Inspection ___/___/_____   Inspector(s) _____

Condition Codes (0=Worst → 9 = Best ; See Chapter 4 for Ratings Schedules)

Upper Plenum (if present)	Rating	**Roadway**	Rating
Underside of Roof	____	Underside of Ceiling/Roof Slab	____
Top of Ceiling Slab	____	Top of Invert	____
Right Wall (if applicable)	____	Right Wall	____
Left Wall (if applicable)	____	Left Wall	____
Lower Plenum (if present)		**Miscellaneous Appurtenances**	
Underside of Invert	____	Safety Walks	____
Bottom of Plenum	____	Railings	____
Right Wall (if applicable)	____	Utility/CCTV Supports	____
Left Wall (if applicable)	____	Finishes (Check one) Excellent o Good o Fair o Poor o	

Comments

_____
_____
_____
_____
_____

TUNNEL AGENCY NAME
HIGHWAY TUNNEL FIELD INSPECTION FORM
SUPPLEMENTAL TUNNEL SEGMENT SKETCHES

_____
_____

General Information

Tunnel Name _____

Begin Station _____ End Station _____

                                                       Or

                                                       Panel Number _____

Year Built _____ Liner Type _____

Date of Inspection ____/____/_____ Inspector(s)_____

_____

THIS SPACE TO BE USED
FOR SUPPLEMENTAL SKETCHES,
AND/OR COMMENTS

TUNNEL AGENCY NAME
HIGHWAY TUNNEL FIELD INSPECTION FORM
TUNNEL SEGMENT PHOTO LOG SHEET

General Information

Tunnel Name _____

Begin Station _____ End Station _____

                                      Or

                                      Panel Number _____

Year Built _____ Liner Type _____

Date of Inspection ____/____/_____ Inspector(s)_____

Roll No.	Counter No.	Description

TUNNEL AGENCY NAME
RAIL TRANSIT TUNNEL FIELD INSPECTION FORM
TUNNEL SEGMENT CONDITION CODES

General Information

Line _____   Track _____

Name of Station Ahead _____

Name of Station Behind _____

Begin Station _____   End Station _____

Or

Panel Number _____

Year Built _____   Liner Type _____

Date of Inspection ___/___/_____   Inspector(s) _____

Condition Codes   (0=Worst → 9 = Best ; See Chapter 4 for Ratings Schedules)

Track Area	Rating	**Miscellaneous Appurtenances**	Rating
Underside of Roof	___	Safety Walks	___
Concrete Track Supports	___	Railings	___
Right Wall	___	Utility Supports	___
Left Wall	___		

Comments

_____
_____
_____
_____
_____
_____

TUNNEL AGENCY NAME
RAIL TRANSIT TUNNEL FIELD INSPECTION FORM
SUPPLEMENTAL TUNNEL SEGMENT SKETCHES

General Information

Line _____   Track _____

Name of Station Ahead _____

Name of Station Behind _____

Begin Station _____   End Station _____

Or

Panel Number _____

Year Built _____   Liner Type _____

Date of Inspection ___/___/_____   Inspector(s) _____

THIS SPACE TO BE USED
FOR SUPPLEMENTAL SKETCHES,
AND/OR COMMENTS

TUNNEL AGENCY NAME
RAIL TRANSIT TUNNEL FIELD INSPECTION FORM
TUNNEL SEGMENT PHOTO LOG SHEET

General Information

Line _____   Track _____

Name of Station Ahead _____

Name of Station Behind _____

Begin Station _____ End Station _____

Or

Panel Number _____

Year Built _____   Liner Type _____

Date of Inspection ___/___/_____   Inspector(s) _____

Roll No.	Counter No.	Description

E. METHODS OF ACCESS

To access the various structural elements for up-close visual inspection requires that additional equipment be used. A man-lift truck, a rail-mounted vehicle, ladders, and/or removable scaffolding is required.

These types of equipment will permit the inspectors to gain an up-close, hands-on view of most of the structural elements. Binoculars can also be used to locate surface defects from nearby manlifts or ladders, where up-close access is difficult to achieve. It is preferred, however, that up-close, non-destructive testing be used on all tunnel surfaces.

F. SAFETY PRACTICES

Another significant duty of the inspection team is to ensure that safety practices are followed at all times. Along with the safety of inspection personnel, the inspection teams should use caution when inspecting to prevent danger to the public, to tunnel personnel, and to members of the inspection team. If possible, it is best to have the owner close the tunnel when inspections are being conducted. Also, all inspectors should wear reflective vests. Other safety practices that are more specific to highway and transit tunnels are discussed below.

1. Highway

For highway tunnels, appropriate protective devices and vehicles should be properly positioned when using lifts and ladders, and where deemed necessary to warn that equipment is in the roadway. All traffic control devices should conform to and be positioned as directed by the *Manual on Uniform Traffic Control Devices (MUTCD)*. This document is endorsed by the Federal Highway Administration (FHWA) and is published by the American Traffic Safety Services Association (ATSSA), the Institute of Transportation Engineers (ITE), and the American Association of State Highway and Transportation Officials (AASHTO). Also, the *Manual for Work Zone Traffic Control Devices* should be reviewed. It should be noted that some local jurisdictions might require the presence of the local or state police in the event of a tunnel closure or partial closure. This would be to help ensure the safety of the inspection team and the motorist that typically would be using the tunnel.

2. Rail Transit

For rail transit tunnels, inspectors should take extra precaution to avoid contact with the third rail or catenary system. If possible, the third rail or catenary system should be de-energized in the area where the inspection is taking place. This can be accomplished during pre-planning coordination with the tunnel owner. Lock-outs and tags must be used to engage the switch to ensure that the power does not inadvertently re-energize while the inspection is occurring. If the power cannot be de-energized, it is suggested that the tunnel owner provide appropriate personnel to remain with the inspection team during the inspection period. In addition, most rail transit agencies will require that a flagman be present when conducting inspections in the track area.

CHAPTER 4:
INSPECTION PROCEDURES – GENERAL DISCUSSION

A. INSPECTION OF CIVIL/STRUCTURAL ELEMENTS

1. Frequency

The tunnel owner should establish the frequency for up-close inspections of the tunnel structure based on the age and condition of the tunnel. For new tunnels, this time period could be as great as five years. For older tunnels, a much more frequent inspection time period may be required, possibly every two years. This up-close inspection is in addition to daily, weekly, or monthly walk-through general inspections.

2. What to Look For

This section provides the procedures for inspections as well as the definitions of defects common to concrete, steel, masonry, and timber structures. The identification of structural defects will be accomplished via both visual inspection and non-destructive techniques.

The visual inspection must be made on all exposed surfaces of the structural elements. All noted defects should be measured and documented for location. Severe spalls in the concrete surface should be measured in length, width, and depth. Severe cracks should be measured in length and width. Corrosion on steel members should be measured for the length, width, and depth of the corrosion. The inspectors should clear away debris, efflorescence, corrosion, or other foreign substances from the surfaces of the structural element prior to performing the inspection. Once the defect is noted, it should be classified as minor, moderate, or severe as explained in the following sections.

Particular attention should be paid to determining if differential settlement has occurred in transition areas of the tunnel. Transition areas are those in which the tunnel support conditions change, such as between sections of rock and soil tunneling or between the tunnel and ventilation or station buildings. The location of these areas should be evident from any existing as-built drawings. Differential settlement is often the cause of other defects, which is why extra time should be spent investigation these transition areas.

In addition to visual inspection, structural elements should be periodically sounded with hammers to identify defects hidden from the naked eye. As a result of a hammer strike on the surface, the structural element will produce a sound that indicates if a hidden defect exists. A high-pitched sound or a ringing sound from the blow indicates good material below the surface. Conversely, a dull thud or hollow sound indicates a defect exists below the surface. Such a defect in concrete may signify a delamination is present or that the concrete is loose and could spall off. A hollow sound in timber may indicate advanced decay. Once the defect is found, the surface in the vicinity of the defect should be tapped until the extent of the affected area is determined. This procedure is to be applied to concrete and timber surfaces but should also be used on steel especially where corrosion is evident.

For concrete or masonry surfaces that are accessible, a non-destructive, ultrasonic test method such as "Impact-Echo" may be utilized. Impact-Echo is an acoustic method that can determine locations and extent of flaws/deteriorations, voids, debonding of re-bars, thickness of concrete. The use of this method helps to mitigate the need for major retrofit since the deterioration can be detected at an early stage and repairs performed.

It should be noted that in the 1960s some tunnel owners began to develop maximum allowable rates of water infiltration to be used as a guide to determine original design and subsequent repairs if the amount of infiltration increases. One such owner was the Bay Area Rapid Transit (BART) system in California; they set a limit of 0.8 liters/minute per 75 linear meters (0.2 gpm per 250 linear feet) of tunnel. This translates to 3 liters/minute per 300 linear meters (0.8 gpm per 1000 linear feet) of tunnel. Some tunnel owners have adopted this criteria while still others may use a limit of 3.8 liters/ minute per 300 linear meters (1 gpm per 1000 linear feet) of tunnel. These limits are for reference purposes only, with the main emphasis for determining repair needs placed on the location of the leak and the condition of the tunnel components that are affected. For this reason, minor, moderate, and severe leakage rates are given in the following sections for use in classifying individual leaks. Also, it is important for the tunnel inspector to be aware of any tunnel waterproofing system that was installed during the construction of the tunnel. Many of the common civil/structural defects are listed below:

a) <u>Concrete Structures (Refer to ACI 201.1R-92 for representative pictures of these defects)</u>

(1) <u>Scaling</u>

The gradual and continuing loss of surface mortar and aggregate over an area classified as follows:

- Minor Scale – Loss of surface mortar up to 6 mm (¼ in) deep, with surface exposure of coarse aggregates.

- Moderate Scale – Loss of surface mortar from 6 mm (¼ in) to 25 mm (1 in) deep, with some added mortar loss between the coarse aggregates.

- Severe Scale – Loss of coarse aggregate particles as well as surface mortar and the mortar surrounding the aggregates. Depth of loss exceeds 25 mm (1 in).

(2) **Cracking**

A crack is a linear fracture in the concrete caused by tensile forces exceeding the tensile strength of the concrete. Cracks can occur during curing (non-structural shrinkage cracks) or thereafter from external load (structural cracks). They may extend partially or completely through the concrete member. Cracks are categorized as follows:

- Transverse Cracks – These are fairly straight cracks that are roughly perpendicular to the span direction of the concrete member. They vary in width, length, and spacing. These cracks may extend completely through the slab or beam as well as through curbs and walls supporting the safety walk.

- Longitudinal Cracks – These are fairly straight cracks that run parallel to the span of the concrete slab or beam. They vary in width, length, and spacing. The cracks may extend partially or completely through the slab or beam.

- Horizontal Cracks – These cracks generally occur in walls but may exist on the sides of beams where either encased flanges or reinforcement steel have corroded. They are similar in nature to transverse cracks.

- Vertical Cracks – Vertical cracks occur in walls and are similar to longitudinal cracks in slabs and beams.

- Diagonal Cracks – These cracks are roughly parallel to each other in slabs and are skewed relative to the centerline of the structure. They are usually shallow and are of varying length, width, and spacing. When found in the vertical faces of beams, they signify that a potentially serious problem exists.

- Pattern or Map Cracks – These interconnected cracks vary in size and form networks similar to that of sun cracking observed in dry areas. They vary in width from barely visible, fine cracks to well-defined openings. They are found in both slabs and walls.

- D-Cracks – These cracks are a series of fine cracks at rather close intervals with random patterns.

- Random Cracks – These are meandering irregular cracks on the surface of concrete. They have no particular form and do not logically fall into any of the classifications described above.

All cracks in non-prestressed members may be classified as follows:

- Minor - Up to 0.80 mm (0.03 in).

- Moderate - Between 0.80 mm (0.03 in) and 3.20 mm (0.125 in).

- Severe - Over 3.20 mm (0.125 in).

Any crack over 0.10 mm (0.003 in) in a prestressed member should be classified as severe. Any crack ≤ 0.10 mm (0.003 in) should be classified as moderate.

(3) Spalling

Spalling is a roughly circular or oval depression in the concrete. It is caused by the separation and removal of a portion of the surface concrete revealing a fracture roughly parallel, or slightly inclined, to the surface. Usually, a portion of the depression rim is perpendicular to the surface. Often reinforcement steel is exposed. Spalling may be classified as follows:

- Minor – Less than 12 mm (½ in) deep or 75 mm (3 in) to 150 mm (6 in) in diameter.

- Moderate – 12 mm (½ in) to 25 mm (1 in) deep or approximately 150 mm (6 in) in diameter.

- Severe – More than 25 mm (1 in) deep and greater than 150 mm (6 in) in diameter and any spall in which reinforcing steel is exposed.

(4) Joint Spall

This is an elongated depression along an expansion, contraction, or construction joint. This defect should be classified as described above.

(5) Pop-Outs

These are conical fragments that break out of the surface of the concrete leaving small holes. Generally, a shattered aggregate particle will be found at the bottom of the hole, with a part of the fragment still adhering to the small end of the pop-out cone.

- Minor – Leaving holes up to 10 mm (0.40 in) in diameter, or equivalent.

- Moderate – Leaving holes between 10 mm (0.40 in) and 50 mm (2 in) in diameter, or equivalent.

- Severe – Leaving holes 50 mm to 75 mm (2.0 in to 3.0 in) in diameter, or equivalent. Pop-outs larger than 75 mm (3 in) in diameter are spalls.

(6) Mudballs

These are small holes that are left in the surface by the dissolution of clay balls or soft shale particles. Mudballs should be classified in the same way as pop-outs.

(7) Efflorescence

This is a combination of calcium carbonate leached out of the cement paste and other recrystalized carbonate and chloride compounds, which form on the concrete surface.

(8) Staining

Staining is a discoloration of the concrete surface caused by the passing of dissolved materials through cracks and deposited on the surface when the water emerges and evaporates. Staining can be of any color although brown staining may signify the corrosion of underlying reinforcement steel.

(9) Hollow Area

This is an area of a concrete surface that produces a hollow sound when struck by a hammer. It is often referred to as delaminated concrete.

(10) Honeycomb

This is an area of a concrete surface that was not completely filled with concrete during the initial construction. The shape of the aggregate is visible giving the defect a honeycomb appearance.

(11) <u>Leakage</u>

This occurs on a region on the concrete surface where water is penetrating through the concrete.

- Minor – The concrete surface is wet although there are no drips.
- Moderate – Active flows at a volume less than 30 drips/minute.
- Severe – Active flows at a volume greater than 30 drips/minute.

b) <u>Steel Structures</u>

(1) <u>Corrosion</u>

Corroded steel varies in color from dark red to dark brown. Initially, corrosion is fine grained, but as it progresses, it becomes flaky or scaly in character. Eventually, corrosion causes pitting in the member. All locations, characteristics, and extent of the corroded areas should be noted. The depth of severe pitting should be measured and the size of any perforation caused by corrosion should be recorded. Corrosion may be classified as follows:

- Minor – A light, loose corrosion formation pitting the paint surface.
- Moderate – A looser corrosion formation with scales or flakes forming. Definite areas of corrosion are discernible.
- Severe – A heavy, stratified corrosion or corrosion scale with pitting of the metal surface. This corrosion condition eventually culminates in loss of steel section and generally occurs where there is water infiltration.

(2) <u>Cracks</u>

Cracks in the steel may vary from hairline thickness to sufficient width to transmit light through the member. Any type of crack is serious and should be reported at once. Look for cracks radiating from cuts, notches, and welds. All cracks in the steel will be classified as severe.

(3) <u>Buckles and Kinks</u>

Buckles and kinks develop mostly because of damage arising from thermal strain, overload, or added load conditions. The latter condition is caused by the failure or the yielding of adjacent members or components. Erection or collision damage may also cause buckles, kinks, and cuts.

(4) Leakage

This occurs on a region of the steel surface where water is penetrating through a joint or crack.

- Minor – The steel surface is wet although there are no drips.

- Moderate – Active flows at a volume less than 30 drips/minute

- Severe – Active flows at a volume greater than 30 drips/ minute.

(5) Protection System

Steel is generally protected by a paint system, by galvanizing, or by the use of weathering steel. Most existing structures use either paint or galvanized steel. Paint systems fail through peeling, cracking, corrosion pimples, and excessive chalking. The classification of the degree of paint system deterioration is tied to both the physical condition of the paint and the amount of corrosion of the member as follows:

- Minor – General signs of deterioration of the paint system but no corrosion yet present.
- Moderate – Paint generally in poor condition and corrosion is present but not serious. (No section loss.)

- Severe – Paint system has failed and there is extensive corrosion and/or section loss.

c) Masonry Structures

(1) Masonry Units

The individual stones, bricks, or blocks should be checked for displaced, cracked, broken crushed, or missing units. For some types of masonry, surface deterioration or weathering can also be a problem.
- Minor – Surface deterioration at isolated locations. Minor cracking.

- Moderate – Slight dislocation of masonry units; large areas of surface scaling.

- Severe – Individual masonry units significantly displaced or missing.

(2) <u>Mortar</u>

The condition of the mortar should be checked to insure that it is still holding strongly. It is particularly important to note cracked, deteriorated, or missing mortar if other deterioration is present such as missing or displaced masonry units.

- Minor – Shallow mortar deterioration at isolated locations.

- Moderate – Mortar generally deteriorated, loose, or missing mortar at isolated locations, infiltration staining apparent.

- Severe – Extensive areas of missing mortar; infiltration causing misalignment of tunnel.

(3) <u>Shape</u>

Masonry arches act primarily in compression. Flattened curvature, bulges in walls, or other shape deformations may indicate unstable soil conditions.

(4) <u>Alignment</u>

The vertical and horizontal alignment of the tunnel should be checked visually.

(5) <u>Leakage</u>

A region on the masonry surface where water is penetrating through a joint or crack.

- Minor – The masonry surface is wet although there are no drips.

- Moderate – Active flows at a volume less than 30 drips/minute.

- Severe – Active flows at a volume greater than 30 drips/minute.

d) <u>Timber Structures</u>

(1) <u>Decay</u>

Decay is the primary cause of timber deterioration and is caused by living fungi, which feed on the cell walls of timber. Molds, stains, soft rot (least severe), and brown or white rot (most severe) are common types of fungi that cause decay. Timber may become discolored and soft and section loss may occur. Any decay should be noted and the amount of section loss should be recorded. Decay may be classified as follows:

- Minor – Discoloration of timber, molds, or stains present

- Moderate – Timber surface soft with section loss less than 15 percent.

- Severe – Brown or white rot present with section loss greater than 15 percent.

(2) <u>Insects</u>

Any presence of insect infestation should be noted and type of insect recorded, if known. Saw dust or powdered dust on or around the timber member may also indicate the presence of insects and should be noted. Termites and carpenter ants are common types of insects that cause timber deterioration.

(3) <u>Checks/Splits</u>

Checks are cracks in timber, which extend partially through the timber member and are caused by shrinkage due to drying or seasoning of the timber. Cracks that extend completely through the member are called splits. All checks should be noted along with the percentage of penetration through the member. Checks may be classified as follows:

- Minor – Surface checks perpendicular to the plane of stress or isolated checks parallel to the plane of stress.

- Moderate – Checks with less than 15 percent penetration into the timber perpendicular to the plane of stress or isolated checks with less than 40 percent penetration parallel to the plane of stress.

- Severe – Checks greater than 15 percent penetration into the timber perpendicular to the plane of stress or numerous checks greater than 40 percent penetration parallel to the plane of stress.

(4) Fire Damage

Classification is based on the amount of fire damage.

- Minor – Black or charred surface. No appreciable section loss.
- Moderate – Less than 15 percent section loss.
- Severe – Greater than 15 percent section loss.

(5) Hollow Area

A hollow area indicates advanced decay in the interior of a timber member or the presence of insects. All hollow areas should be noted as to size and location.

(6) Leakage

A region on the timber surface where water is penetrating through a joint, check/split, or the timber itself.

- Minor – The timber surface is wet although there are no drips.
- Moderate – Active flows at a volume less than 30 drips/minute.
- Severe – Active flows at a volume greater than 30 drips/minute.

e) Connection Materials

(1) Bolts

The connection bolts on fabricated concrete, steel, and cast iron liners may be discolored due to moisture and humidity conditions in the tunnel. This condition does not downgrade the structural capacity of the bolt. Particular attention should be given to bolts in regions of leakage to ensure that no detrimental loss of section has occurred. If losses in section are observed, such bolts should be noted for replacement. Also, the location of all missing or loose bolts should be noted.

- Minor – Bolts are discolored, but have no section loss.
- Moderate – Bolts are deteriorated with up to 15 percent section loss.

- Severe – Bolts are deteriorated with greater than 15 percent section loss. However, bolts with deterioration approaching 50 percent or more should be replaced.

 (2) Gaskets

Gaskets between segmental tunnel liners can be lead, mastic, or rubber. These gaskets can become dislodged from the joint due to infiltrating water or loosening of the joint bolts. They also can fail due to chemical or biological deterioration of the material caused by the infiltrated water. Structural movements of the liner can also tear or otherwise distort the gasket and cause it to leak. All gasket deficiencies should be noted as to extent and location.

3. Safety – Critical Repairs

The inspection may reveal severe defects that could pose danger to the traveling public, tunnel personnel, or inspection team members. When this occurs, this particular severe defect should be categorized for a "critical repair." This categorization would deem that one of the following critical actions be taken:

- Close the tunnel until the severe defect is removed or repaired if such a defect is accessible by vehicles or trains

- Cordon off the area from public access until the defect can be removed or repaired

- Shore up the structural member if this is appropriate.

It is imperative that the inspection team coordinates with the tunnel owner in advance and be prepared to take the "critical action." Oftentimes, this type of action is required for delaminated concrete that is on the verge of falling. The inspection team, tunnel personnel, or a specialty contractor could possibly perform the removal. This activity can be very dangerous and all safety precautions should be taken to prevent injury to inspectors and repair personnel.

4. Condition Codes

Elements will be rated using the guidelines explained below. A numerical rating of 0 to 9 will be assigned to each structural element, 0 being the worst condition and 9 being the best condition. This rating system is a modified form of the one described in the *Bridge Inspector's Training Manual* published by the FHWA. A general description of the rating system is shown below in Table 4.1 and in Table 4.2 this system is specifically related to different liner types.

If a tunnel owner desires to use this system for mechanical or electrical systems or other tunnel appurtenances, then the codes can be adapted to represent a smaller set of conditions. An example would be to give numbers to the following conditions: excellent, good, fair, poor, and serious. This may be done in order to track conditions through the use of a tunnel management software program.

Rating	Description
9	Newly completed construction.
8	Excellent condition - No defects found.
7	Good condition - No repairs necessary. Isolated defects found.
6	Shading between "5" and "7."
5	Fair condition - Minor repairs required but element is functioning as originally designed. Minor, moderate, and isolated severe defects are present but with no significant section loss.
4	Shading between "3" and "5."
3	Poor condition - Major repairs are required and element is not functioning as originally designed. Severe defects are present.
2	Serious condition - Major repairs required immediately to keep structure open to highway or rail transit traffic.
1	Critical condition - Immediate closure required. Study should be performed to determine the feasibility of repairing the structure.
0	Critical condition - Structure is closed and beyond repair.

Table 4.1 – General Condition Codes

The rating is dependent upon the amount, type, size, and location of defects found on the structural element as well as the extent to which the element retains its original structural capacity. To judge the extent to which the structural element retains its original structural capacity, the inspector must understand how the element is designed and how the defect affects this design.

To aid the inspectors in evaluating specific conditions for the various kinds of tunnel elements, more specific guidelines are presented hereafter to ensure consistency in assigning condition codes.

5. **Tunnel Segments**

   a) Cut-and-Cover Concrete Box Tunnels and Concrete/Shotcrete Inner Liners

   For several highway tunnels and for station areas in rail transit tunnels, the concrete/shotcrete surfaces may be covered with another finish material as described under Tunnel Finishes in Chapter 2, Section A, Part 5. For ceramic tile and epoxy finishes, the general condition of the underlying concrete surfaces is to be evaluated and assessed a condition rating based upon the cracks and leakage through the finish material.

In addition to the descriptions of potential defects presented in Chapter 4, Section A, Part 2, the concrete/shotcrete elements should be rated according to the Condition Code Summary in Table 4.2.

The inspector needs to use good engineering judgment when assessing the overall condition rating of the segment being inspected. Although the specific guidelines presented in Table 4.2 are an excellent tool to ensure consistency of evaluations among different inspectors, the defects presented will not always fall into the categories. For example, if a segment of the cut-and-cover concrete tunnel shows no defects (i.e., no delaminations, no spalls, and no exposed reinforcement steel) other than one severe crack with severe active leakage, the inspector may still want to rate this segment in "poor condition" because a major repair is required to restore the element to good condition.

b) <u>Soft-Ground Tunnel Liners</u>

These liners include fabricated steel, precast concrete, cast iron or masonry liners, as well as connection bolts and gaskets on the fabricated liners.

In addition to the descriptions of potential defects presented in Chapter 4, Section A, Part 2, the inspector should be aware of the following requirements for these liners:

- The ends of precast concrete liners may have an embedded steel plate across the full width of the liner plus steel plate inserts for bolting two end-to-end liners together. The condition of the embedded steel plate is synonymous with the precast liner and therefore should be inspected for degree of corrosion.

- The connection bolts on fabricated concrete, steel, and cast iron liners may be discolored due to moisture and humidity conditions in the tunnel. This condition does not downgrade the structural capacity of the bolt. Particular attention should be given to bolts in regions of leakage to ensure that no detrimental loss of section has occurred. If losses in section are observed, such bolts should be noted for replacement.

- The tunnels should be generally observed for uniform cross-sectional shape from radial soil pressures. As a means of monitoring possible changes in cross section, measurements should be taken at approximately 60 m (200 ft) intervals on the inside face of the liners between spring lines and from the underside of the ceiling/roof at 12:00 to the top of rail for rail transit tunnels or to top of walkway for highway tunnels. Yellow paint should be used to identify the measurement locations.

    The ratings for the elements of the tunnel liners are based on the general condition codes in Table 4.1 and for degree of deterioration as shown in Table 4.2.

c) <u>Rock Tunnel Liners</u>

These include cast-in-place concrete and shotcrete liners.

The entire exposed portion of the tunnel liner above the roadway slab or rail transit invert slab should be inspected for typical concrete deficiencies described in Chapter 4, Section A, Part 2. In addition, the lining should be generally observed for uniform cross-sectional shape. As a means of monitoring possible changes in the cross section, measurements should be taken at approximately 60 m (200 ft) intervals between the spring line or vertical sidewalls and from the underside of the ceiling/roof at tunnel centerline to the top of the walkway for highway tunnels or the top of rail for rail transit tunnels. Yellow paint should be used to mark the measurement locations.

The ratings for the rock tunnel, concrete/shotcrete linings are based on the rating scale and descriptions given at the beginning of this section in Table 4.1 for degree of deterioration and as supplemented in Table 4.2.

d) <u>Timber Liners</u>

The exposed portion of the timber liner shall be inspected for typical timber deficiencies described in Chapter 4, Section A, Part 2. The ratings for the timber liners are based on the rating scale and descriptions given at the beginning of this section in Table 4.1 for degree of deterioration and as supplemented in Table 4.2.

e) <u>Track Supports</u>

The track supports in this section are for direct fixation fasteners to concrete plinth pads, continuous concrete pedestals, and the invert slab.

These concrete elements should be inspected for the deficiencies described in Chapter 3, Section A, Part 2 and rated according to the general criteria listed at the beginning of this section.

f) <u>Finishes</u>

As described previously, the primary tunnel finishes include ceramic tiles, porcelain-enameled metal panels, precast concrete panels, and epoxy coatings.

Whereas it does not make sense to use a rating scale of 0 to 9 for such finishes, the inspector could use ratings such as excellent, good, fair and poor condition that follow the general definitions given at the beginning of this section.

Table 4.2 - Condition Code Summary

Rating	General Description	Cut-and-Cover Box Tunnels and Concrete/Shotcrete Inner Liners	Soft-Ground Tunnel Liners	Rock Tunnel Liners	Timber Liners
9	Newly completed construction	Newly completed construction	Newly completed construction	Newly completed construction	Newly completed construction
8	**Excellent condition** – No defects found	**Excellent condition** – No defects found	**Excellent condition** – No defects found	**Excellent condition** – No defects found	**Excellent condition** – No defects found
7	**Good Condition** – No repairs necessary. Isolated defects found	**Good condition** – No repairs necessary although certain elements contain isolated minor deficiencies and minor presence of efflorescence. No delaminations or spalls are present	**Good condition** – No repairs necessary. Steel, precast concrete, cast iron, and masonry liners have isolated minor defects. Steel plates, shapes, and liners have isolated locations of minor surface corrosion but with no section loss. Precast concrete liners and safety walk panels contain not more than one minor crack. Masonry exhibits a minor presence of efflorescence with only minor cracks at greater than 3 m (10 ft) intervals. Connection bolts are discolored, and minor leakage is occurring through the gaskets between liners	**Good condition** – No repair necessary. Concrete/shotcrete liner contains minor circumferential cracks at greater than 3m (10 ft) intervals with a minor presence of efflorescence	**Good condition** – No repair necessary. Timber exhibits isolated locations of minor checks, minor decay, and minor water leakage
6	Shading between "5" and "7"	Shading between "5" and "7"	Shading between "5" and "7"	Shading between "5" and "7"	Shading between "5" and "7"
5	**Fair condition** – Minor repairs required but element is functioning as originally designed. Minor, moderate, and isolated sever defects are present but with no significant section loss	**Fair condition** – Minor repairs required but element is functioning as originally designed. Concrete elements contain moderate cracks at 1.5 m (5 ft) to 3 m (10 ft) intervals with moderate presence of efflorescence and minor to moderate active leakage. Minor delaminations, spalls, map cracking, and staining exist on the concrete but no reinforcement steel is exposed	**Fair condition** – Minor repairs required but liner is functioning as originally designed. Steel, precast concrete, cast iron, and masonry liners have numerous minor defects. Steel plates, shapes, and liners are surface corroded throughout but with no significant section loss. Precast concrete liners and safety walk panels have moderate spalls and more than two minor cracks. Masonry contains moderate cracks at 1.5 m (5 ft) to 3 m (10 ft) intervals with moderate presence of efflorescence and minor to moderate active leakage at isolated locations. Connection bolts (not more than one for every two liner segments) require replacement or retightening. Moderate leakage is present between liners	**Fair condition** – Minor repairs are required but element is functioning as originally designed. Concrete/shotcrete lining contains moderate circumferential cracks at 1.5 m (5 ft) to 3 m (10 ft) intervals, not more than one longitudinal moderate crack, moderate presence of efflorescence, and minor to moderate active leakage. Minor delaminations, spalls, map cracking and staining are present but no reinforcement is exposed	**Fair condition** – minor repairs required but liner is functioning as originally designed. Timber exhibits numerous minor defects and isolated moderate decay, checks, and moderate leakage. Isolated timber members contain splits
4	Shading between "3" and "5"	Shading between "3" and "5"	Shading between "3" and "5"	Shading between "3" and "5"	Shading between "3" and "5"

Table 4.2 - Condition Code Summary *(continued)*

Rating	General Description	Cut-and-Cover Box Tunnels and Concrete/Shotcrete Inner Liners	Soft-Ground Tunnel Liners	Rock Tunnel Liners	Timber Liners
3	**Poor condition** – Major repairs are required and element is not functioning as originally designed Severe defects are present	**Poor condition** – Major repairs are required and element is not functioning as originally designed Concrete elements contain numerous moderate cracks with extensive efflorescence, severe leakage, and staining Delaminations and spalls are present over 50% of the concrete surface and exposed reinforcement steel has up to 15% section loss	**Poor condition** – Major repairs are required The precast, steel, and cast iron liner elements exhibit extensive severe deterioration such that the liners can no longer achieve the full original design capacity, although still retaining some degree of their load-carrying capacity Masonry contains numerous moderate cracks with extensive efflorescence, leakage, and staining Delaminations of the masonry layers, slight dislocation of certain masonry units, and loose or missing mortar are prevalent at isolated locations Severe active leakage is occurring through the masonry or between adjacent liner segments at isolated locations Connection bolts are deteriorated with up to 15% section loss and are loose or missing at isolated locations	**Poor condition** – Major repairs are required and element is not functioning as originally designed Concrete/shotcrete lining has extensive longitudinal and circumferential cracks with extensive efflorescence, leakage, and staining Delaminations and spalls are present over 50% of the lining surface and exposed reinforcement steel has up to 15% section loss	**Poor condition** – Major repairs are required and liner is not functioning as originally designed Liner elements have numerous moderate defects of decay, checks, splits, and leakage over 50% of the liner area
2	**Serious condition** – Major repairs required immediately to keep structure open to highway or rail transit traffic	**Serious condition** – Major repairs required immediately to keep structure open to highway or rail transit traffic Concrete elements contain extensive severe cracks, delaminations, spalls, and leakage Exposed reinforcement steel has up to 40% section loss	**Serious condition** – Major repairs required immediately to keep structure open to highway or rail transit traffic The liner elements have extensive, serious material deterioration and are severely deflected such that the elements can no longer support the design loads without immediate repairs Masonry contains extensive severe cracks, delaminations, and missing masonry units Severe active leakage is occurring at numerous locations within the tunnel segment Connection bolts are deteriorated with up to 50% section loss and are loose or missing at several locations	**Serious condition** – Major repairs required immediately to keep structure open to highway or rail transit traffic Concrete/shotcrete lining has extensive severe cracks, delaminations, spalls, and active leakage Exposed reinforcement steel has up to 40% section loss	**Serious condition** – Major repairs required immediately to keep structure open to highway or rail transit traffic Timber contains extensive severe decay, checks, splits, and leakage over 50% of the timber surface Numerous timber members are completely deteriorated
1	**Critical condition** – Immediate closure required Study should be performed to determine the feasibility of repairing the structure	**Critical condition** – Immediate closure required Study should be performed to determine the feasibility of repairing the structure	**Critical condition** – Immediate closure required The liner elements have major deterioration and have lost all capacity to sustain the original design loadings The masonry is severely deteriorated such that major sections are missing and can no longer support the design loading Connection bolts are deteriorated in excess of 50% and are loose or missing at several locations	**Critical condition** – Lining has extensively cracked and deflected significantly The lining has lost all capacity to sustain design loads	**Critical condition** – Lining has extensively cracked and deflected significantly The lining has lost all capacity to sustain design loads
0	**Critical condition** – Structure is closed and beyond repair	**Critical condition** – Structure is closed and beyond repair	**Critical condition** – Structure is closed and beyond repair	**Critical condition** – Structure is closed and beyond repair	**Critical condition** – Structure is closed and beyond repair

For ceramic tiles, the inspector should lightly tap a majority of tiles to determine if they are loose from the substrate. Such tapping will easily reveal a hollow sound if the tiles have been dislodged. Where hollow areas are noted, the inspector should freely define the extent of the hollow area by tapping all adjacent tiles. The area should be so noted in the inspection findings. In addition, the inspector should evaluate the tiles for reflectivity and general condition (broken or missing).

For porcelain-enameled metal panels and precast concrete panels, it is crucial that the attachments to the substrate be carefully evaluated and assessed, as a loose panel would pose a dangerous condition. For metal panels, any corrosion of the panels along the edges or at scrapes/nicks, will indicate a downgrading of the surface condition. For precast concrete panels, careful attention should be given to determine if there are any cracks or delaminated areas in the panels.

Epoxy coatings are not structural in nature, but do offer a protective covering to the concrete for changing environmental conditions or for additional reflectivity to improve visibility in the tunnel. Note the condition of the coating primarily due to peeling or debonding from the substrate.

g) Drainage Systems

It is crucial that the tunnel drainage system be inspected to ensure that it is capable of handling the water flow for which it was designed. This water flow could be ground water from the exterior of the tunnel lining, rain water that enters the portals or vent structures, and drainage from emergency fire protection systems during their use. Failure of this system could result in ponding of the water and subsequent flooding of the tunnel space, thus posing a safety hazard to the tunnel occupants. Transit tunnels are particularly sensitive to this due to the amount of equipment that can reside on the tunnel invert such as the signals and communications equipment, and the third rail power systems. Excessive ponding of water around such equipment can lead to equipment malfunctioning and potential shut down of the system.

Similarly to the tunnel finishes above, it is recommended that the inspector use a rating scale of excellent, good, fair, and poor condition. The rating should be based on the particular system component's ability to convey or store the required amount of water, i.e., whether it is clogged or leaking.

h) Miscellaneous Tunnel Appurtenances

These appurtenances include railings, safety walks, utility supports, CCTV camera supports, enclosures adjacent to the roadway/track etc.

These miscellaneous elements are to be carefully inspected for the degree of deterioration in the elements as well as the extent to which they retain their original design structural capacity. The inspector shall assign a rating for each element based upon the general ratings listed at the beginning of this section and in Table 4.1.

B.   INSPECTION OF MECHANICAL SYSTEMS

1.   Frequency

It should be noted that if a proper preventive maintenance program is strictly adhered to, the main purpose of an in-depth inspection is to verify that the mechanical systems are performing as expected. For this reason, recommended preventive maintenance procedures and frequencies for specific mechanical systems and equipment are listed in the accompanying *Maintenance and Rehabilitation Manual*. Table 4.3 lists the inspection frequencies for highway and rail transit agencies throughout the country for their mechanical systems, which include pumps, fans, motors, etc. Also, a percentage of all the tunnel owners that inspect their mechanical systems are given for each specific frequency.

Type of Agency	Daily	Weekly	Twice/Month	Monthly	3 Months	6 Months	12 Months	24 Months	36 Months	84 Months	120 Months
Highway	X	X		X	X	X	X	X	X	X	X
	4.0%	4.0%		28.0%	4.0%	4.0%	20.0%	24.0%	4.0%	4.0%	4.0%
Rail Transit	X		X	X	X	X		X			
	8.3%		8.3%	50.0%	8.3%	16.7%		8.3%			

Table 4.3 – Mechanical Inspection Frequency

The inspection frequencies shown should only be used as a guide to what is currently done, not as a suggestion of what should be done. It is up to the tunnel owner to determine the frequency of these in-depth inspections. They can be performed concurrently with the civil/structural inspections or as deemed necessary by the owner because of the age of the mechanical equipment and the amount of equipment needed for proper tunnel operation. It is difficult to determine from the data provided in Table 4.3 the actual scope of the inspection being conducted. It can be assumed that a frequent inspection (generally less than semi-annually, but dependent on the equipment being inspected) is typically a walk-through visual inspection in which the inspector is only looking at critical mechanical equipment to verify that it is functioning properly. On the other hand, a less frequent inspection (generally semi-annually or greater) should be an indicator of a full, in-depth inspection in which every piece of mechanical equipment is inspected and its condition assessed to ensure proper operation.

2. What to Look For

The mechanical inspection will consist of verifying the condition and operation of tunnel equipment and systems. The inspection will include a review of the physical condition of each piece of equipment for damage due to environmental and operational conditions. Any procedures involving the operation of system components must be coordinated with the tunnel owner prior to performing any testing. Each system or piece of equipment should be checked for operation, unless operation of the equipment would cause damage to equipment and/or inspection personnel, or significant disruption to the operation of the tunnel. Any equipment that cannot be operated should be identified, its physical condition noted, and such information immediately reported to the tunnel owner. The inspection should encompass the following systems: tunnel ventilation, air conditioning, heating, controls, plumbing, tunnel drainage, fire protection, and wells / septic systems. Each system should be inspected as follows:

a) Tunnel Ventilation

The inspection of the tunnel ventilation system should include, as a minimum, the following items:

- Review the maintenance records for each piece of equipment and note any special or frequent previous maintenance problems.
- Note the physical condition of each fan, airway, louver, motor-operated dampers, and drive trains.
- Verify that each fan and the associated motor-operated dampers and components are operational.
- Engage a special testing firm to perform vibration analysis on the fans, motors, and bearings during typical fan operations and inspect the fan drive system and bearings.
- Ensure that the airways, where accessible, are free of obstruction and debris.
- Test the operation of the CO monitoring equipment (if such a system exists in highway tunnels).

b) Air Conditioning

The inspection of the air conditioning systems in control rooms, etc., should include the following items:

- Review the maintenance records for each piece of equipment and note any special or frequent previous maintenance problems.
- Note the physical condition of air handling units, condensing units, packaged units, chillers, pumps, cooling towers, exposed air distribution systems, cooling piping, and terminal units.

- Verify that the system is operational. The temperatures at the time of the inspection may dictate if the system is able to be in operation.
- Engage a special testing firm to perform vibration analysis and inspections on chillers, cooling towers, and pumps.
- Engage a special testing firm to perform lube oil analysis on all bearing lubricants.

c) <u>Heating</u>

The inspection of the support area heating system should include the following items:

- Review the maintenance records for each piece of equipment and note any special or frequent previous maintenance problems.
- Note the physical condition of air handling units, pumps, steam and water distribution systems, terminal units, boilers, exposed air distribution systems, heating piping, and steam converters.
- Engage a special testing firm, preferably one that is commissioned by the National Board of Boiler and Pressure Vessel Inspectors, to analyze the boilers to determine their operating efficiency and to inspect the boiler breeching for corrosion and holes.
- Verify that the system is operational. The temperatures at the time of the inspection may dictate if the system is able to be in operation.

d) <u>Controls</u>

The inspection of the tunnel controls should include a visual observation that the control panel indicators represent the operating condition(s) of the equipment each control serves.

The use of a SCADA (Supervisory Control and Data Acquisition) System often controls the entire facility. These systems operate with a minimal amount of hardware maintenance, with the exception of the component level sensors. Software changes for additional programming and periodic upgrades are required to maintain flexibility and reliability of system operation.

e) <u>Plumbing</u>

The inspection of the support area plumbing system should be conducted according to any applicable plumbing code requirements and should also include the following:

- Review the maintenance records for the plumbing system and note any special or frequent maintenance problems.
- Note the physical condition of the bathroom fixtures, water heaters, and drainage system.
- Verify that the plumbing fixtures are operational and the piping is free of leakage.
- Look for watermarks on tunnel surfaces to identify locations of leaks in plumbing system.

f) Tunnel Drainage

The tunnel drainage system including sump pumps should be inspected to determine if the tunnel drains are clear of debris to permit water runoff to flow freely through the drains.

g) Fire protection

The inspection of the fire protection system should include the following items:

- Review the maintenance/inspection records for the system and note any unusual maintenance issues.
- Note the physical condition of the fire protection system in the tunnel and tunnel support areas. This will include the fire extinguishers, hose connections, pumping systems, piping, circulating pumps, and hose reels.
- Note the physical condition of the fire protection storage tanks, alarms, and level switches.
- Check fire control panel for faulty detectors, signals, and wiring.

C. INSPECTION OF ELECTRICAL SYSTEMS

1. Frequency

As with the mechanical systems, it should be noted that if a proper preventive maintenance program is strictly adhered to, the main purpose of an in-depth inspection is to verify that the electrical systems are performing as expected. For this reason, recommended preventive maintenance procedures and frequencies for specific electrical systems and equipment are listed in the accompanying *Maintenance and Rehabilitation Manual*. Table 4.4 lists the inspection frequencies for highway and rail transit agencies throughout the country for inspecting electrical systems. Also, the percentage of tunnel owners that inspect their electrical systems is given for each specific frequency.

Table 4.4 – Electrical Inspection Frequency

Type of Agency	Daily	Weekly	Monthly	3 Months	6 Months	12 Months	24 Months	36 Months	84 Months	120 Months
Highway	X	X	X	X	X	X	X	X	X	X
	3.1%	3.1%	15.6%	3.1%	3.1%	28.1%	34.4%	3.1%	3.1%	3.1%
Rail Transit		X	X		X	X	X			
		15.4%	61.5%		7.7%	7.7%	7.7%			

These frequencies should only be used as a guide to what is currently done, not what should be done. It is up to the tunnel owner to determine the frequency of and which items should be checked during these in-depth inspections. They can be performed concurrently with the civil/structural inspections or as deemed necessary by the owner because of the age of the electrical equipment and the amount of equipment needed for proper tunnel operation. It is difficult to determine from the data provided in Table 4.4 the actual scope of the inspection being conducted. It can be assumed that a frequent inspection (generally less than semi-annually, but dependent on the equipment being inspected) is typically a walk-through visual inspection in which the inspector is only looking at critical electrical equipment to verify that they are functioning properly. On the other hand, a less frequent inspection (generally semi-annually or greater) should be an indicator of a full, in-depth inspection in which every piece of electrical equipment is inspected and its condition assessed to ensure proper operation. Further recommendations on inspections are given in NETA MTS 1 and NFPA 70B.

2. **What to Look For**

The electrical system inspection will consist of verifying the condition and operation of all of the following systems: power distribution, emergency power, lighting, fire detection, and communication. Each of these systems are described herein and are to be inspected for the specific requirements listed below and the following general items:

- Visibly inspect wiring systems for damage and corrosion.
- Ensure that all enclosures and box covers are in place and secure.
- Check for conformity to NFPA 70, 70B, 70E, 72, 130, and NETA MTS 1.
- Check that all disconnects are properly identified as to the items they disconnect.
- Check that all loads are properly identified as to the source or means of disconnect.
- For all large power systems, Electrical Safety Operating Diagrams should be posted to comply with OSHA and NFPA 70E.

a) <u>Power Distribution System</u>

This system consists of the electrical equipment, wiring, conduit, and cable used for distributing electrical energy from the utility supply (service entrance) to the line terminals of utilization equipment. The system would include equipment such as transformers, switchgear, switchboards, unit substations, panelboards, motor control centers, starters, switches, and receptacles. Specific inspection includes:

- Take voltage and load readings on the electrical system using any of the installed meters.
- Check that all indicator gages on the transformers show that fluid levels, temperatures, and pressures are within range.
- Check for signs of damage and overheating of all equipment.
- Check that adequate working space is provided in accordance with NFPA 70, Article 110 and is clear in front of equipment with no material stored in the working space.
- Evaluate the condition of enclosures and conduit and ensure that all enclosures and box covers are in place, conduits are not broken, etc.
- Visibly inspect wiring systems for damage and corrosion.
- Check power distribution system for conformity to NFPA 70 and NFPA 130.
- Check that all disconnects are properly identified as to what items they disconnect.
- Check that all loads are properly identified as to the source or means of disconnects.
- Check all motor controllers for proper operation.
- Have a NETA testing agency perform a thermographic (infrared) inspection for hot spots and an internal inspection, and note any deficiencies. Have this same agency review the previous maintenance records to see if prior discrepancies were corrected. Verify that all tests meet industry standards, including NETA MTS1.

b) <u>Emergency Power System</u>

This system consists of the electrical equipment, wiring, conduit, and cable used for providing electrical power in case of utility service failure. Equipment included in this system consists of emergency generators or uninterruptible power supply (UPS) systems, transfer switches, and other equipment supplying emergency power.

- Ascertain the ability of the emergency power system to operate when the normal power fails, by disabling the normal power supply (i.e., the supply that supplies any transfer switch or other means of transferring loads) and operating the emergency system with selected emergency loads for a sufficient period to evaluate its condition.

- Have a NETA testing agency perform an internal inspection and an inspection for hot spots, and note any deficiencies. Have this same agency review the previous maintenance records to see if prior discrepancies were corrected. Verify that all tests meet industry standards, to include NETA MTS1 and NFPA110.

c) <u>Lighting System</u>

This system consists of the electrical equipment, wiring, conduit, cable, luminaries, sensors, and controllers used to provide lighting for the tunnel.

- Measure the light levels within highway tunnels using an Illuminating Engineering Society (IES) LM-50 device and compare the results against the requirements of IES RP-22.
- Measure the light levels at intervals suggested by IES LM-50.
- Measure the light levels at emergency egress exits and compare with the IES Handbook recommendations.
- Inspect all lighting that is above the roadway surfaces for visible damage, to include corroded or damaged housings, loose attachments, broken lenses, and burnt out bulbs. Also, note if lenses should be cleaned.
- Verify the operation of the lighting controls for the different ranges of nighttime and daylight illumination.
- For transit tunnels, verify that all emergency and continual-use lighting is operational and provides the required amount of illumination.
- For transit tunnels, test operation of Emergency Trip Switch (ETS) lighting, which is linked to the third rail power system to indicate when third rail power is properly shut off. These lights are typically spaced every 240 m (800 ft).

d) <u>Fire Detection System</u>

This system consists of control panels, initiating devices (heat and smoke detectors, pull-stations, etc.), notification appliances (strobes, horns, etc.), wiring, conduit, and cable used to detect a fire in the tunnel.

- Inspect the fire detection system by operating the drill switch and assuring that all of the annunciators and notification appliances operate.
- Check existing records to determine if the system has been tested at regular intervals in accordance with NFPA 72. NFPA 72 requires that a copy of the records for the last seven years be available.

e) Communication System

This system consists of the communication equipment (SCADA, CCTV cameras, telephones, radios, etc.) used to provide communication within and from the tunnel.

- Verify the operation of the SCADA system by ensuring a positive indication is received for each required operation.
- Verify that the CCTV cameras, telephones, radios, or other communication devices are operational.
- Inspect traffic signals for proper operation during all phases.
- Verify that any over-height detectors are not triggering at any heights just below the desired setting and also verify that they are triggering at or just above the desired setting.

D. INSPECTION OF OTHER SYSTEMS/APPURTENANCES

1. Inspection of Track Elements

It is recommended that rail transit owners require their internal inspectors, outside consultants, or specialized testing agencies be familiar with and follow the recommended procedures established by their own internal guidelines or the current revision of the US DOT's Federal Railroad Administration – Office of Safety's, Code of Federal Regulations for Title 49, *Track Safety Standards Part 213 Subpart A to F, Class of Track 1-5 (TSS Part 213)*. The current revision of this document is dated June 9th, 2001. In fact, it is recommended that each inspection team be required to have a copy of this "pocket-size" book and any owner's specific requirements available when conducting track inspections.

TSS Part 213 provides the following guidelines for class of track and maximum operating speeds for passenger trains shown in Table 4.5.

Class of Track	Maximum Allowable Operating Speed for Passenger Trains, km/h (mph)
Class 1	25 (15)
Class 2	50 (30)
Class 3	100 (60)
Class 4	130 (80)
Class 5	140 (90)

Table 4.5 – Passenger Train Operating Speeds

It is not the intent of this manual to duplicate the material in TSS Part 213. Rather, key elements of track will be described and the general deficiencies that an inspector must look for and evaluate when conducting track inspections will be listed.

a) <u>Frequency</u>

TSS Part 213 Subpart F describes the frequencies for inspecting various classifications of track. It also explains how such inspections can be performed by vehicle or on foot. Inspection frequencies vary from twice weekly for typical rail transit Class 4 and 5 track to weekly for lower speed track (Classes 1 through 3). TSS Part 213 Subpart F also provides inspection frequency guidelines for switches, track crossings and lift rail assemblies or other transition devices on moveable bridges. These requirements include monthly visual inspections on foot and the physical operation of mechanized switches every three months. In addition, TSS Part 213 requires that the rail on all Class 3 and higher track be inspected for rail or rail joint defects with specialized rail defect detection equipment at least once a year.

The questionnaires sent out revealed that rail transit owners perform track inspections that vary from daily to yearly based upon their internal procedures. The established frequencies need not change unless the requirements of TSS Part 213 are not being achieved.

b) <u>What to Look For</u>

This section provides general guidance on defects that occur in track elements as listed in TSS Part 213, Subparts C, D, and E. Specifically, these Subparts cover the following track elements:

- Subpart C – Track Geometry (Gage, Alignment, Curves – Elevation and Speed Limitations, Elevations of Curved Track – Runoff, and Track Surface).
- Subpart D – Track Structure (Ballast, Crossties, Gage Restraint Measurement Systems, Defective Rails, Rail End Mismatch, Continuous Welded Rail, Rail Joints, Torch Cut Rail, Tie Plates, Rail Fastening Systems, Turnouts and Track Crossings, Switches, Frogs, Spring Rail Frogs, Self-Guarded Frogs, and Frog Guide Rails and Guard Faces – Gage).
- Subpart E – Track Appliances and Track-Related Devices (Derails).

An inspection program should ensure that each of the track elements in each Subpart are reviewed and evaluated to maintain safe operation of the trains. A brief description of the inspection required for some of the major track elements is as follows:

(1) <u>Rail</u>

Inspect the rail for horizontal and vertical cracks in the steel, horizontal and vertical split heads, transverse and compound fissures, fractures, split web, piped rail, bolt-hole crack, head web separation, broken base, detail fracture, engine burn fracture, broken or defective weld, and surface defects. See TSS Part 213 and the *Rail Defect Manual* compiled by Sperry Rail Service for explanation of these defects. The severity of these defects range from minor to major.

(2) Gage

The gage of the track is the distance from the center to center of rail heads measured at right angles to the rails in a plane five-eighths of an inch below the top of the rail head. The gage may deviate from construction over time, thus exceeding tolerances for tangent and curved tracks. TSS Part 213 gives the appropriate gage distance according to classification of track as follows in Table 4.6:

Class of Track	Gage must be at least	But not more than
Class 1	1400 mm (4' – 8")	1450 mm (4' – 10")
Class 2 and 3	1400 mm (4' – 8")	1444 mm (4' – 9 ¾")
Class 4 and 5	1400 mm (4' – 8")	1438 mm (4' – 9 ½")

Table 4.6 – Track Gage Distances

(3) Alignment

Alignment of tracks may not deviate from uniformity more than the amounts shown in Table 4.7:

Class of Track	Tangent Track	Curved Track	
	The deviation of the mid-off-set from a 18.6 m (62 ft) line[1] may not be more than ___ mm (inches)	The deviation of the mid-ordinate from a 9.3 m (31 ft) chord[2] may not be more than ___ mm (inches)	The deviation of the mid-ordinate from a 18.6 m (62 ft) chord[2] may not be more than ___ mm (inches)
Class 1	125 (5)	N/A[3]	125 (5)
Class 2	75 (3)	N/A[3]	75 (3)
Class 3	43 (1 ¾)	43 (1 ¾)	43 (1 ¾)
Class 4	37 (1 ½)	25 (1)	37 (1 ½)
Class 5	18 (¾)	12 (½)	15 (⅝)

Table 4.7 – Track Alignment

[1] The ends of the line shall be at points on the gage side of the line rail, 15 mm (⅝ in) below the top of the railhead. Either rail may be used as the line rail; however, the same rail shall be used for the full length of that tangential segment of track.

[2] The ends of the chord shall be at points on the gage side of the outer rail, 15 mm (⅝ in) below the top of the railhead.

[3] N/A – Not Applicable

(4) <u>Curves</u>

The maximum crosslevel on the outside rail of a curve may not be more than 200 mm (8 in) on Track Classes 1 and 2 and 175 mm (7 in) on Classes 3 through 5.

(5) <u>Fasteners/Bolts/Spikes</u>

Each of these elements is extremely important for safe operation of the rail transit system, especially in securing the gage of the track. Spikes should be inspected to ensure they are tight and snug against the rail if they are used for lateral restraint of the rail. Fasteners include special clips or attachments on both regular jointed and continuous welded rail (CWR) and should be inspected for missing, broken, or loose fasteners. For CWR, fasteners can also serve as longitudinal restraint.

Special testing of the fasteners is required to ensure the proper clamping force is being applied to restrain the rail. Bolts are used at rail joints for splicing the rail or for attaching tie plates to the underlying anchorage system, whether ties or direct fixation. They should be inspected for condition and missing, loose, or broken fasteners.

(6) <u>Tie Plates</u>

Tie plates are used to distribute the rail uniformly to the supporting tie or concrete bearing system with grout. These should be inspected for condition and to determine if uniform bearing is being achieved. If point loading on the tie plates is prevalent, this may affect the gage of the rail, alignment, or curvature.

(7) <u>Crossties</u>

Crossties support and secure the rail. They are typically made of timber or precast concrete although fiber reinforced plastic ties are currently being manufactured or used to rehabilitate existing timber ties. They should be inspected for condition and for effective support in accordance with TSS Part 213, Subpart D, Section 213.109.

Timber crossties should be inspected to insure that they are not: broken through, split or otherwise impaired such that ballast can work through or spikes and rail fasteners cannot be held, so deteriorated that the tie plate or base of rail can move laterally 12 mm (½ in) relative to the crosstie, or cut by the tie plate through more than 40 percent of the crosstie's thickness.

For concrete crossties, they should be inspected for cracks, deteriorated concrete, spalls, etc.

(8) Ballast

Inspect the ballast for condition. Insufficient or fouled ballast is described as any ballast that will not a) transmit and distribute the load of the track and railroad rolling equipment to the subgrade; b) restrain the track laterally, longitudinally, and vertically under dynamic loads imposed by railroad rolling equipment or thermal conditions; c) provide adequate drainage for the track; and d) maintain proper track crosslevel, surface, and alignment.

(9) Rail Joints

Rail joints occur where the ends of two rails meet and are spliced together with bolts to maintain vertical integrity of the rail. These joints should be inspected for cracks or loose joint bars; worn joint bars (such that excessive vertical movement is permitted); broken joint bars; and missing, broken, or deteriorated bolts. The proper number of crossties in the vicinity of the rail joint is also critical and should be reviewed in accordance with crossties mentioned before.

2. Inspection of Power Systems (Third Rail/Catenary)

a) Third Rail Power System

(1) Frequency

It is recommended that visual inspections of the third rail system be performed for normal tunnel sections or at crossovers in tunnels on a monthly basis. This visual inspection should be made on foot and should note any defects that would affect the movements of the electric rail transit vehicles. In addition, it is recommended that testing equipment such as ohmmeters; a direct current power supply, generator, or equivalent alternating current supply; and miscellaneous leads be used to test resistance of rail joints on a yearly basis. Any deficiencies should be repaired during this testing period.

(2) What to Look For

The third rail system provides power to electric rail transit vehicles via direct contact with the third rail from current collectors (shoes) attached to the transit vehicles. The third rail system is comprised of the steel contact rail, protection boards, protection board brackets, insulators, insulator caps, anchors, and negative running rail bonded joints as shown in Figure 2.19. A brief description of the inspection required for some of the major third rail system elements is as follows:

(a) Steel Contact Rail

The contact rail should be inspected to determine if it is resting evenly and uniformly on all insulators, for excessive wear and damage on the contact surface, to determine if its alignment follows the same radius as any curve in the tunnel, and to ensure that all ends are terminated with end approach castings.

(b) Contact Rail Insulators

The contact rail should be inspected to verify that insulators are present, that they rest directly on each bracket, and that they are held in place by a centering cup that forms an integral part of the bracket. Each insulator shall be covered with an insulator cap. This cap is held in place by the lug hole, which is an integral part of the insulator. Inspect each insulator for condition. Those that are dirty should be thoroughly cleaned; any that are broken or chipped should be replaced.

(c) Protection Board

Protection boards should be of sufficient length to be supported by not less than two brackets. On curves these boards should be cut to conform to the radius of the curve. Inspect these boards to ensure they are in good condition, properly attached, and cover the contact rail.

(d) Protection Board Brackets

For timber ties, inspect the protection board brackets to determine if the brackets are placed on the long timber ties, that they are horizontally gaged accurately, that they rest directly on the tie, and that they are fastened by two lag screws. Also, verify that no brackets are installed on ties supporting joints in the running rail. For concrete base supports, the brackets are to be fastened with bolts and a 3 mm (1/8 in) thick polyethylene pad is to be placed between the steel bracket and the concrete for isolation purposes.

(e) Contact Rail Splices

All contact rail splice joints, except joints at end approaches, should have a bonded joint. Two bonds are required at each bonded joint, one on each side of the contact rail. Inspect this bonded joint by performing a resistance test on a two-foot length across the joint with a length of solid rail necessary to give an equal resistance. The solid rail reading should not be greater than 800 mm (32 in). Use the following testing equipment to perform the test:

- Ductor, low resistance ohmmeter.
- Direct Current (D.C.) power supply.
- Miscellaneous leads.
- Generator or equal for Alternating Current (A.C.) supply.

(f) <u>Negative Running Rail Bonded Joints</u>

Inspect all negative running rail bonds, impedance bond locations, turnout and crossover bonds to complete a continuous negative circuit. Ensure that no defective joints exist. Perform a resistance test on a 750 mm (30 in) length of rail across the joint with a length of solid rail necessary to give an equal resistance. The solid rail reading should not be greater than 1200 mm (4 ft). Use the same equipment as described for the contact rail splice test.

(g) <u>Third Rail Insulated Anchor Arms</u>

Inspect all bolts, insulators, contact clamps, and anchor plates used at contact rail anchor locations. Inspection should note loose or worn bolts, nuts or clamps, broken insulators, or any removal of assemblies.

b) <u>Catenary Power System</u>

(1) <u>Frequency</u>

It is recommended that the catenary inspection include two levels – a visual inspection and an in-depth inspection. Two agencies have specific frequency requirements that may be indicative of other rail transit agencies and are discussed herein. These agencies include Metro-North Commuter Railroad and AMTRAK.

Metro-North recommends that a visual inspection of the catenary system be made on foot at bi-monthly intervals. They also recommend that an in-depth inspection be made from a track vehicle with a high-level platform bi-annually. On the other hand, AMTRAK recommends that a visual inspection of the catenary system be made from the head end of a train on a weekly basis. AMTRAK also requires a quarterly geometry car inspection, a yearly catenary car inspection, and a bi-annual up-close inspection and repair be made from the top of a catenary car, wire train, or highway-rail vehicle.

(2) <u>What to Look For</u>

There are at least two documents available that give specific requirements for inspection of the catenary system. These include the Catenary System Inspection Procedures for the Metro-North Commuter Railroad and the Catenary Inspection

Manual for AMTRAK. Since Metro-North was included in the inventory for this project, some of their recommended procedures will be briefly defined herein. However, this would not preclude an owner from using his own or even AMTRAK's recommended procedures. A brief description of the inspection required for some of the major catenary system elements is given below. Note that a particular element may have different procedures for a visual inspection than for an in-depth inspection.

(a) <u>Visual Inspection</u>

In addition to noting general observations for the following elements, a walk-through inspection should record excessive arcing in a matrix that shows location, train speeds, number and type of pantographs, direction of travel, climatic conditions, and any unusual circumstances. This will help determine if arcing is due to equipment mal-adjustment or to a pattern of circumstances.

- Support and Registration Insulators – Check for broken sheds and any build-ip of deposits that could cause tracking. Insulator support steel should be checked for missing or loose nuts and any evidence of movement.

- Hangers – Check that alignment is vertical. Hangers consistently leaning one direction indicate that stretch or slippage has occurred and this should be investigated immediately. Make note of detached hangers and determine cause; this may be due to loose or damaged carbons on pantographs. Check contact wire clip for evidence of impact.

- Jumpers – Check "C" jumpers and full section continuity jumpers at overlaps for loose clamps, movement or evidence of burning and that they do not sag below contact wire level.

- Pull-Off Arrangements – Check for evidence of clamp slippage and ensure that heel settings are higher than the contact wire and that the drop bracket is vertical. Verify that the messenger is positioned vertically over the contact wire at pull-off locations.

- Anchors – Check to see if the anchors supporting the catenary system within the tunnel are in good condition and anchored firmly to the substrate. Note any deficiencies.

(b) In-depth Inspection

The in-depth inspection will be used to perform the following inspection tasks as well as give inspection personnel the opportunity to complete general repairs and preventive maintenance.

- Contact Wire Wear – Check contact wire wear at each registration point with particular attention to phase gaps and overlaps. Verify that vertical thickness of contact wire does not measure less than 11 mm (0.42 in). If this is not the case, then the contact wire should be replaced in that location.

- Clamped Electrical Connectors – Randomly remove and check clamped connections at "C" jumpers, feeder points and full section overlap jumpers for corrosion or burning. If the above conditions are found, the clamps should be removed, cleaned, and tightened. In addition, high melting point grease should be applied to stranded conductors.

- Hangers – Check for evidence of mechanical wear or electrical arcing. Inspect neoprene sleeves between the messenger and the retainer.

- Messenger Supports – Check for electrical tracking across the insulator and check stainless steel wire and thimbles for signs of mechanical wear; replace as necessary.

- Registration Assembly – Check registration components at same location as messenger assembly for wear. Open contact wire clamps, check for wear, and regrease. Inspect hinge pin, clevis pin and all bolted connections. Tighten or replace as necessary.

- Support and Registration Insulators – Check for contamination, signs of electrical tracking, and broken or chipped sheds. Check tightness of fixing bolts into ends of insulators.

- Overlaps – Verify the contact wire profiles at overlaps to ensure efficient transition of the pantographs. Inspect underside of contact wires for signs of arcing and adjust, if necessary.

- Section Insulators – Check for evidence of burning of the skids and arcing horns. Adjust turnbuckles on the support hangers of each unit to keep units level and vertically in line with contact wire.

- Disconnect Switches – Open and close to ensure operation of switches.

3. Inspection of Signal/Communication Systems

Similarly to track inspections, it is recommended for inspection of signal systems that rail transit owners require their internal inspectors, outside consultants, or specialized testing agencies be familiar with and follow the recommended procedures established by their own internal guidelines or the US DOT's Federal Railroad Administration – Office of Safety's, Code of Federal Regulations for Title 49, Part 236 – *Rules, Standards, and Instructions Governing the Installation, Inspection, Maintenance, and Repair of Signal and Train Control Systems, Devices, and Appliances*[8], further referred to as FRA Part 236. For communication systems, the same office of the Federal Railroad Administration has produced Title 49, Part 220 – *Railroad Communications*[9], further referred to as FRA Part 220. However, this publication contains little in regard to inspection and testing and more on actual operations of such systems.

It should be noted that there is no current requirement for rail transit tunnel owners to follow the guidelines in FRA Part 236 or Part 220, but in lieu of developing their own, many transit agencies have adopted these as standards.

Given the complexity of FRA Part 236, it will not be reproduced in this manual in its entirety, but rather key sections have been identified and generalized to give basic direction for inspection of signal systems. To establish a comprehensive signal system inspection program, it is recommended that a copy of FRA Part 236 be consulted directly.

    a) <u>Frequency</u>

        (1) <u>Signal Systems</u>

Specific inspection procedures outlined in FRA Part 236 for the various components of the signal system range in frequency from 1 month to 10 years. Some of these frequencies include the following:

Component		Frequency
Switch circuit controller	-	3 months
Switch obstruction	-	monthly
Semaphore/searchlight	-	6 months
Relays (type dependent)	-	2 years
Ground tests (power supplies)	-	3 months
Insulation tests (cables)		
New	-	10 years
Resistance < 500,000 ohms	-	yearly
Timing devices	-	yearly
Interlocking tests	-	2 years
Trip stops		
Height and alignment	-	monthly
Operation	-	6 months

(2) Communication Systems

Specific frequencies for inspection of communication systems range from continuously (by use) to monthly. Examples of these frequencies are shown below:

Emergency telephones	-	monthly
Radiating cables	-	prior to work
Backbone cables	-	continuously(by use)
Communications equipment	-	per manufacturer.

As another source of frequency recommendation, a survey of current rail transit tunnel owners showed actual inspection frequencies for both signal and communication systems that range from daily to every six months, with monthly being most common.

b) What to Look For

(1) Signal Systems

Comprehensive and timely inspections of signal systems are necessary due to the possible consequences if a component of the system fails. According to FRA Part 236: "When any component of a signal system, the proper functioning of which is essential to the safety of train operation, fails to perform its intended signaling function or is not in correspondence with known operating conditions, the cause shall be determined and the faulty component adjusted, repaired or replaced without undue delay." In addition to the visual and operative inspection procedures given in the above frequency section, a brief description of the inspection required for some additional signal system elements is as follows:

- Verify that legible and correct plans are kept at all interlockings, automatic signals, and controlled points.
- Verify that open-wire transmission lines operating at voltage of 750 volts or more shall be placed not less than 1,200 mm (4 ft) above the nearest crossarm carrying signal or communication circuits.
- Verify that each wire is tagged or marked so that it can be identified at each terminal. Also, verify that tags and wires do not interfere with moving parts of any signal apparatus.
- Test the operating characteristics of all parts of semaphores or searchlights.
- If a block signal system is being used, then verify that each signal governing train movements into a block will display its most restrictive aspect when any of the following conditions are met within the block:

- Occupancy by a train, locomotive, or car.
- When points of a switch are not closed in proper position.
- When an independently operated fouling point derail-equipped with switch circuit controller is not in derailing position.
- When a track relay is in de-energized position or a device which functions as a track relay is in its most restrictive state; or when signal control circuit is de-energized.

- Verify that insulated rail joints are maintained to prevent sufficient track circuit current from flowing between the rails separated by the insulation to cause a failure of any track circuit involved.
- Verify that the trip stop arm is maintained at the height above the plane of the tops of the rails and the horizontal distance from its center line to gage side of the nearest running rail are in accordance with the specifications of the carrier.
- Ensure that results of all tests performed are recorded on preprinted or computerized forms provided by the rail transit authority and are retained until the next record is filed or for one year, whichever is greater.

(2) <u>Communication Systems</u>

As with signal systems, it is critical that communication systems be kept in good working condition; therefore, accurate and frequent verification of working condition is pertinent. In addition to the procedures described in the above frequency section, the following actions are suggested:

- Verify that all emergency telephones are in proper operating condition by placing a call from each location.
- Test signal strength of radiating cable.
- Visually inspect backbone cables for signs of degradation.
- Follow all manufacturers' recommendations for inspection and preventive maintenance of communications equipment.

CHAPTER 5:
INSPECTION DOCUMENTATION

A. FIELD DATA

1. Tunnel Structure

The inspection should be thoroughly and accurately documented. For the tunnel structure, the documentation of severe defects should include a sketch showing the location and size of the defect and a verbal description of the defect. All severe defects should be photographed; however, a representative photo of minor or moderate defects will be sufficient. All defects should be described but sketches need only to be made for severe defects.

The sketches of the defect can be made on forms developed during the mobilization phase or on computer screens, as appropriate. These forms should show the necessary plan and elevation views of the structural element to which they pertain. Blank forms should also be provided for additional sketches where deemed necessary by the inspectors. All defects should be located on sketches or the computer screen by dimensioning their location in reference to the beginning or end of the element. Each defect should be dimensioned showing its length, width, and depth (if applicable).

For consistency in documenting the inspection findings, each inspector should use the following system both to describe the defect and to classify them as minor, moderate or severe:

Description of Defect	Classification
Crack - CR	1 - Minor
Scaling - SC	2 - Moderate
Spall - SP	3 - Severe
Staining - ST	
Exposed Reinforcement - E	
Corrosion - C	
Honeycomb - H	
Patch Failure - PF	
Hollow Area - HA	
Debris - D	
Buckle - B	
Efflorescence - EF	
Leakage - LK	
Check - CK	
Rot - RT	
Fire Damage - FD	
Paint Deterioration – PD	

For example, a moderate spall should be labeled as SP2, a severe crack as CR3, etc. This designation should be placed on the sketch and connected to the defect by means of a leader to clearly identify the defect. Samples of completed field sketches for tunnel segments, developed using both a tablet PC data collector and a pre-printed form, are shown in Figures 5.1 and 5.2, respectively. In addition, Figure 5.3 shows an example of a defect location form for an auxiliary space that was completed using a pre-printed form. Upon completion of each section or miscellaneous appurtenance being inspected, a summary rating sheet as shown in Chapter 3, Section D, Part 3, should be completed as a record of the inspection.

Before placing any other information on a form, always complete the spaces at the top of the form identifying the structural element. This will eliminate any confusion when numerous structural elements are completed. A 35 mm or digital camera should be used to take photographs during the inspection. Each inspector will keep a log of all photographs taken. This log should identify the element being inspected, a description of the photograph, the counter number, and the roll number (if appropriate). Examples of photo log sheets and sketch sheets are shown in Chapter 3, Section D, Part 3.

2. **Track Structure**

TSS Part 213, Subpart F, Section 213.241 – Inspection Records requires that rail transit owners are to keep records of each inspection performed. This section identifies when the records are to be completed and the method of record retention – either paper or electronically. Undoubtedly, rail transit owners already have developed in-house forms for these inspections so none will be presented herein. It is suggested that these procedures be followed by all rail transit owners, even if it requires changes to their current documentation procedures.

3. **Specialized Testing Reports**

To inspect certain mechanical, electrical, and other associated track systems requires the use of specialized testing agencies and equipment. All such reports derived from these special testings shall become a part of the documentation of the particular inspection period.

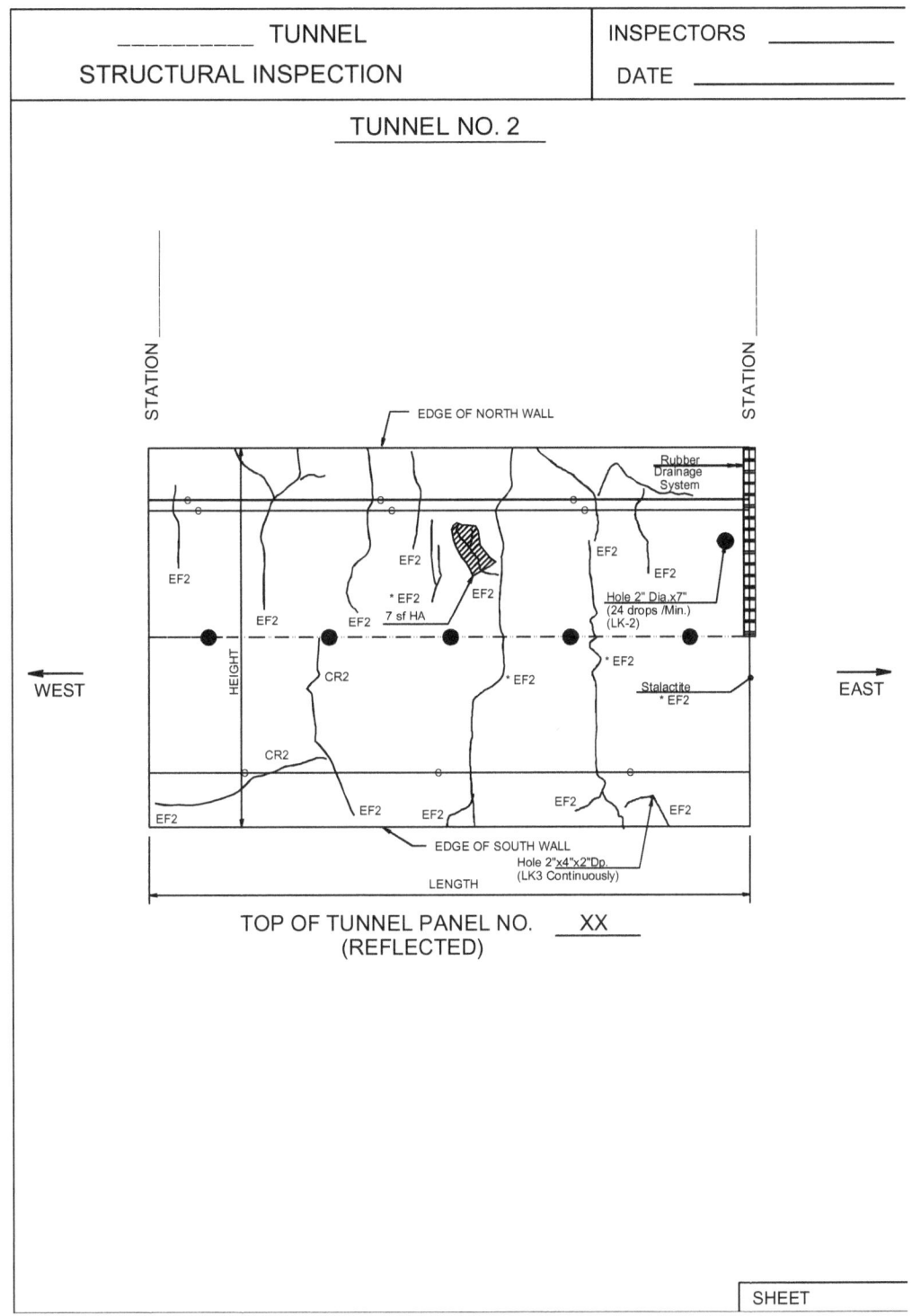

Figure 5.1 – Tunnel Inspection Form (Tablet PC Data Collector)

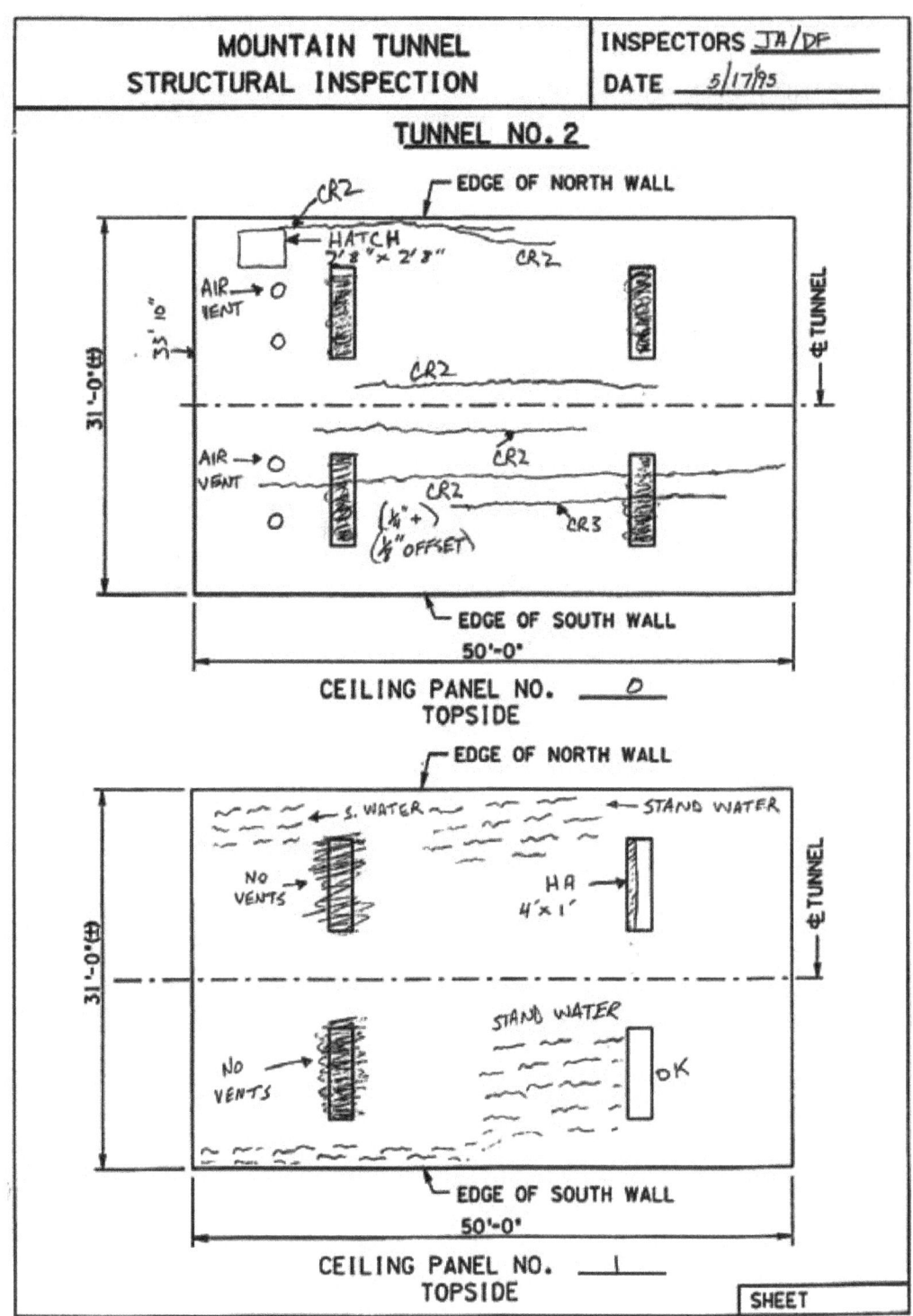

Figure 5.2 – Tunnel Inspection Form (Pre-Printed Form)

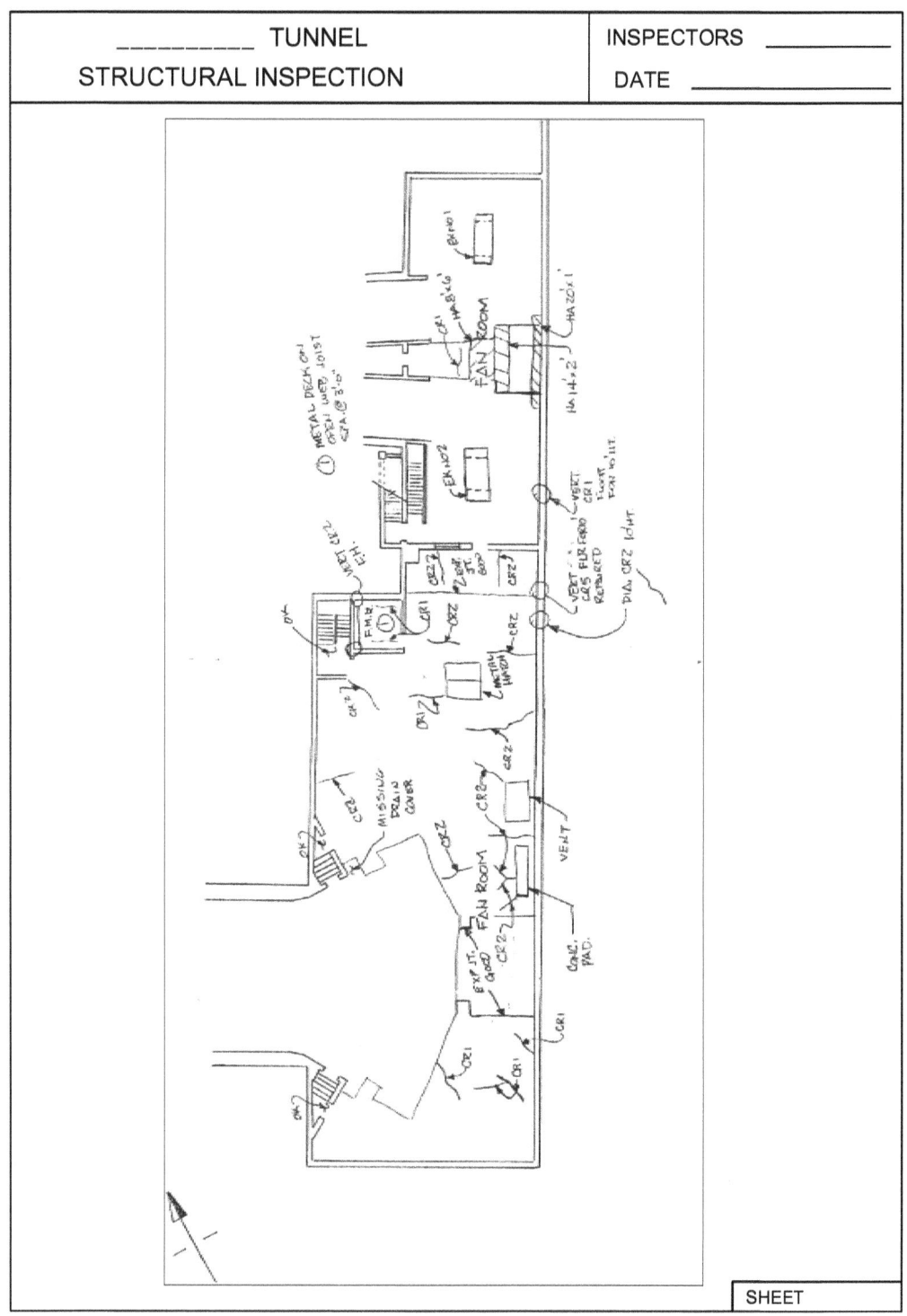

Figure 5.3 –Portal Inspection Form (Pre-Printed Form)

B. REPAIR PRIORITY DEFINITIONS

When summarizing inspection data and making recommendations for future repairs it is necessary to define categories that prioritize the repairs that are to be performed. These recommendations can be included in an inspection report format and/or entered into a structural database for scheduling repairs/rehabilitation and for historical purposes. The following repair classifications are suggested:

1. Critical

As discussed in Chapter 4, Section A, Part 3, a defect requires this designation if it requires "immediate" action including possible closure of the structure or areas affected for safety reasons or from system operation until interim remedial measures can be implemented.

2. Priority

Refers to conditions for which further investigations, design, and implementation of interim or long-term repairs should be undertaken on a priority basis, i.e., taking precedence over all other scheduled work. These repairs will improve the durability and aesthetics of the structure or element and will reduce future maintenance costs. Elements that do not comply with code requirements are also in these classifications. Such repairs should be scheduled for completion within two years.

3. Routine

Refers to conditions requiring further investigation or remedial work that can be undertaken as part of a scheduled maintenance program, other scheduled project, or routine facility maintenance depending on the action required. All items identified in the preventive maintenance program should also be incorporated in this category. Such items should be scheduled for completion after two years.

C. REPORTS

Upon completion of all elements of the inspection, the tunnel owner should require a formal report be developed that summarizes the findings from each element that was inspected. This report will be used to educate the tunnel owner of deficiencies within the tunnel and enable him/her to schedule repairs and allocate sufficient funding.

The report should be supplemented with a computerized database that includes the rating information on structural elements. This database will permit the tunnel owner to query and develop reports as necessary for any inspected element in the tunnel. Below is a suggested outline for the report along with a description of the contents to be included in each section.

- **Letter of Transmittal** – Formal identification of report and introduction to the recipient.

- **Table of Contents** – Self-explanatory.

- **List of Tables** – Used to identify the title and location of any tables that were used to summarize the inspection findings.

- **List of Figures and Drawings** – Used to identify the title and location of any figures or drawings that were used to describe the inspection.

- **List of Photographs** – Used to identify the title and location of any photographs that were taken to document the inspection findings.

- **Executive Summary** – Provide a concise summary of the inspection, findings, and recommended repairs.

- **General Description** – Provide a general description of the tunnel or tunnels that were inspected. This information could include the location of the tunnel(s), age, general geometry, and any other pertinent descriptive information.

- **Inspection Procedures** – The procedures used to perform the inspection of the various tunnel elements below should be explained and illustrated if necessary. Recognition should also be given to any special testing agencies that were used to complete the inspection.

    - Civil/Structural
    - Mechanical
    - Electrical
    - Track, Third Rail, Catenary, Signals, and Communications.

- **Inspection Findings** – A detailed description of the results of the inspection should be included for the various tunnel elements below.

    - Civil/Structural – For civil/structural elements, the report should contain descriptions of the various deficiencies found, their locations and their severity. Any special testing, such as concrete strength, freeze-thaw analysis, or petrographic analysis should be included with the findings.

    - Mechanical – For the mechanical inspections, the general condition and operation of all equipment should be described and deficiencies noted. Specialized testing required to effectively determine the operational condition of the equipment, such as vibration testing and oil analyses, shall be included for historical purposes.

- **Electrical** – For the electrical inspections, the general condition and operation of all equipment should be described and deficiencies noted. Specialized testing required to effectively determine the operational condition of the equipment, such as power distribution and emergency power, shall be included for historical purposes. In addition, measurement of light levels versus that recommended should be provided to the owner. Where testing agencies performed remedial work along with the testing, such as tightening loose wires, etc., it should be included.

- **Track, Third Rail, Catenary, Signals, and Communications** – For the inspection of track, traction power, signals, and communications, the inspectors shall discuss the overall findings and provide copies of specialized testing results.

- **Recommendations** – This section will include actual recommendations for repair/rehabilitation of the tunnel components that were found to be deficient or that did not meet current code requirements. The owner may desire that an estimate of cost be made by the inspectors to correct the defective elements. If substantial rehabilitation is required, the owner may request a life-cycle cost comparison be made comparing repair options in the short-term versus long-term rehabilitation. The repair/rehabilitation should be broken down for each of the main tunnel systems into the different categories listed below, which were defined in the previous section.

    - Critical
    - Priority
    - Routine.

- **Appendices** – The appendices should be used to display detailed and extensive inspection summaries that were too lengthy to include in the body of the report, such as structural panel ratings and lighting illuminance levels. Also, reports provided by special testing agencies should be included in their entirety. Other items that should be included in the appendices are special permits or qualifications that were needed to perform the inspections. An example of this would be confined space entry permits, qualifications, and procedures needed for entering certain areas of a tunnel, such as the air plenums above or below the tunnel space.

This summary report along with the testing results will provide an historical document recording the condition of the tunnel and its inherent systems for the period indicated. This document can be compared to documents from future inspections for tunnel owners to evaluate the overall long term condition of various tunnel elements.

GLOSSARY

AASHTO	-	American Association of State Highway and Transportation Officials
AC	-	Alternating Current
ATSSA	-	American Traffic Safety Services Association
CCTV	-	Closed Circuit Television
Chord	-	A line segment that joins two points on a curve
CO	-	Carbon Monoxide
CWR	-	Continuous Welded Rail
DC	-	Direct Current
ETS	-	Emergency Trip Switch
FHWA	-	Federal Highway Administration
FRA	-	Federal Railroad Administration
FTA	-	Federal Transit Administration
Gunite	-	Term commonly used for fine-aggregate shotcrete
gpm	-	Gallons per minute
IES LM-50	-	Illuminating Engineering Society, Lighting Measurements – 50
IES RP-22	-	Illuminating Engineering Society, Recommended Practices – 22
ITE	-	Institute of Transportation Engineers
Km/h	-	Kilometers per hour
mph	-	Miles per hour
MTS	-	Maintenance Testing Specifications
MUTCD	-	Manual on Uniform Traffic Control Devices

NATM	-	New Austrian Tunneling Method (synonymous with SEM)
NBIS	-	National Bridge Inspection Standards
NBS	-	National Bureau of Standards
NEMA	-	National Electric Manufacturers Association
NETA	-	InterNational Electrical Testing Association
NFPA	-	National Fire Protection Association
OSHA	-	Occupational Safety and Health Administration
PEI	-	Porcelain Enamel Institute
SEM	-	Sequential Excavation Method (synonymous with NATM)
TBM	-	Tunnel Boring Machine
TSS	-	Track Safety Standards
UPS	-	Uninterruptible Power Supply

REFERENCES

Arnoult, J. D., *Culvert Inspection Manual FHWA – IP – 86 – 2*, Federal Highway Administration, 1986.

Bickel, J.; E. King, and T. Kuesel, *Tunnel Engineering Handbook, Second Edition,* Chapman & Hall, New York, 1996.

Haack, A.; J. Schreyer, and G. Jackel, *State-of-the-art of Non-destructive Testing Methods for Determining the State of a Tunnel Lining*, Tunnelling and Underground Space Technology, 10.4 (1995): 413-431.

Metro-North Commuter Railroad, *Manual for Maintenance and Inspection of Constant Tension Catenary Systems*.

National Fire Protection Association, *NFPA 502: Standard for Road Tunnels, Bridges, and Other Limited Access Highways*, 2001.

Sperry Rail Service, *Rail Defect Manual*, Sperry Rail Service, 1968.

SYSTRA Consulting, *Electric Traction Catenary Inspection Field Manual*, AMTRAK, 1998.

U.S. Department of Transportation, Federal Railroad Administration – Office of Safety, *Code of Federal Regulations, Title 49, Track Safety Standards Part 213, Subpart A to F, Class of Track 1-5*, Simmons-Boardman Books, Inc., Omaha, NE, 2001.

U.S. Department of Transportation, Federal Railroad Administration – Office of Safety, *Code of Federal Regulations, Title 49, Part 236, Rules, Standards, and Instructions Governing the Installation, Inspection, Maintenance, and Repair of Signal and Train Control Systems, Devices, and Appliances*, 2001.

U.S. Department of Transportation, Federal Railroad Administration – Office of Safety, *Code of Federal Regulations, Title 49, Part 220, Railroad Communications*, 2001.

www.ingramcontent.com/pod-product-compliance
Lightning Source LLC
Chambersburg PA
CBHW080521110426

42742CB00017B/3191